Telephone Accessories You Can Build

Telephone Accessories You Can Build

Jules H. Gilder

HAYDEN BOOK COMPANY, INC.
Rochelle Park, New Jersey

PATENT NOTICE

Purchasers and other users of this book are advised that projects described herein may be proprietary devices covered by patents owned or applied for and that their inclusion in this book does not, by implication or otherwise, grant any license under such patents or patent rights for commercial use. No one participating in the preparation or publication of this book assumes responsibility for any liability resulting from unlicensed use of information contained herein. Information furnished by the author is believed to be accurate and reliable. However, no responsibility is assumed by the author or publisher for its use.

To my parents who planted the electronic seed and nourished it until it grew into a tree of knowledge.

Library of Congress Cataloging in Publication Data

Gilder, Jules H 1947-
Telephone accessories you can build.

1. Telephone--Apparatus and supplies--Amateurs' manuals. I. Title.
TK9951.G54 621.386 76-48643
ISBN 0-8104-5748-2

Printed in the United States of America

10	11	12	13	14	15	16	17	18	PRINTING
83	84	85	86	87	88	89	90	91	YEAR

Preface

The telephone is probably the most widely used electronic instrument ever invented. This book is designed to familiarize you with the telephone and how it works, and to make it and its accessories a convenience to use. Engineer, technician, and hobbyist alike will find the fifteen projects described both interesting and useful.

The introductory portion of the book covers some basics to supply the background you will need to build the projects presented. Also included is a discussion of some of the regulations concerning telephone systems. This is a very gray area with changes occurring frequently, but the discussion should give you a basic understanding of the situation.

Of particular note is Chapter 3, which gives helpful information on construction techniques and practices. Careful attention to details outlined in Chapter 3 will make construction of the various projects easier and more enjoyable.

Printed circuit layout patterns are provided for all projects which require them. Use of these patterns will simplify construction and minimize the possibility of error. Also presented is a method of printed circuit board fabrication that I have found to be most useful and easy to implement. The last part of Chapter 3 describes a power supply that you will find an important addition to your projects.

Wherever possible a modular approach to project design was taken. More complex projects that appear later in the book call for some of the circuits developed for projects appearing earlier. The sound switch used in the remote ringer project is a good example; it is used in four other projects besides the remote ringer.

The accessories in this book have been designed to work on ordinary private phones of both the dial and pushbutton variety. While some of them will work on party lines, many will not. The projects in the beginning of the book do not require any direct connection to the telephone; you should therefore, encounter no problems with your local telephone company through their use. However, some later projects do require a direct electrical connection to the phone line. Certain telephone companies object to equipment being connected directly to the phone lines, and may therefore require the addition of a company provided coupler. If in doubt, check with your local phone company.

One final point, if you are building projects that connect directly to the phone line, it is best to use a battery supply. This eliminates the possibility of any ac voltage getting onto the line and endangering telephone employees and equipment.

JULES H. GILDER

Brooklyn, N.Y.

Contents

1. Basics

The projects described in this book are designed to be operated with the ordinary house telephone. In order to better understand how these projects work, it will be very helpful if we first look at the basic phone instrument and see how it operates.

One of the more common telephones in use today is the model 500 telephone used in the Bell Telephone System. A picture of the 500-type telephone appears in Figure 1-1 and a typical schematic of it is seen in Figure 1-2.

Sound is picked up by the transmitter of the telephone as variations in pressure caused by a vibrating diaphragm. The diaphragm applies these pressure variations to a variable resistor made of carbon granules. When a battery is connected across the transmitter, the low resistance of the carbon granules allows a current to flow. As the pressure on the granules changes, the resistance of the transmitter changes and thus the current varies in accordance with the pressure of the sound waves on the transmitter. The maximum amount of current allowed to flow is limited by the series resistance of R3.

Sound is reproduced in the telephone by the receiver which contains an electromagnetic earphone. Variations in the electrical current flowing through the receiver cause variations in the magnetic field of the earphone. These variations in the magnetic field cause an iron diaphragm in the earphone to vibrate and thus produce sound.

It's possible to make a telephone with just the receivers and transmitters and some additional switches. But such an instrument would be difficult to use because your voice would be loud in your receiver and low in the receiver of the party you were talking to. The same thing would be true on the other end. To eliminate this, telephone engineers have included devices called varistors in the telephone instrument. Varistors are voltage variable resistors in which the resistance varies inversely with the voltage. So if the voltage across a varistor increases, the resistance decreases and vice versa. This means that the average voltage across the receiver in the phone remains relatively constant. The high voltage produced in your receiver by your microphone is reduced to the same level as the voltage coming from the distant microphone.

In Figure 1-2 varistor RV1 is used to suppress dial pulse clicks in the receiver. The balancing network, composed of varistor RV2, resistor R2, and capacitors C2 and C3 with the windings of the induction coil, form a hybrid arrangement which provides simultaneous two-way operation over a two-wire circuit. Capacitor C1 and resistor R1 make up a dial pulse filter to suppress high-frequency interference to nearby radio receivers. Varistors RV2 and RV3 with R1 also reduce the efficiency of the transmitter on short loops from the

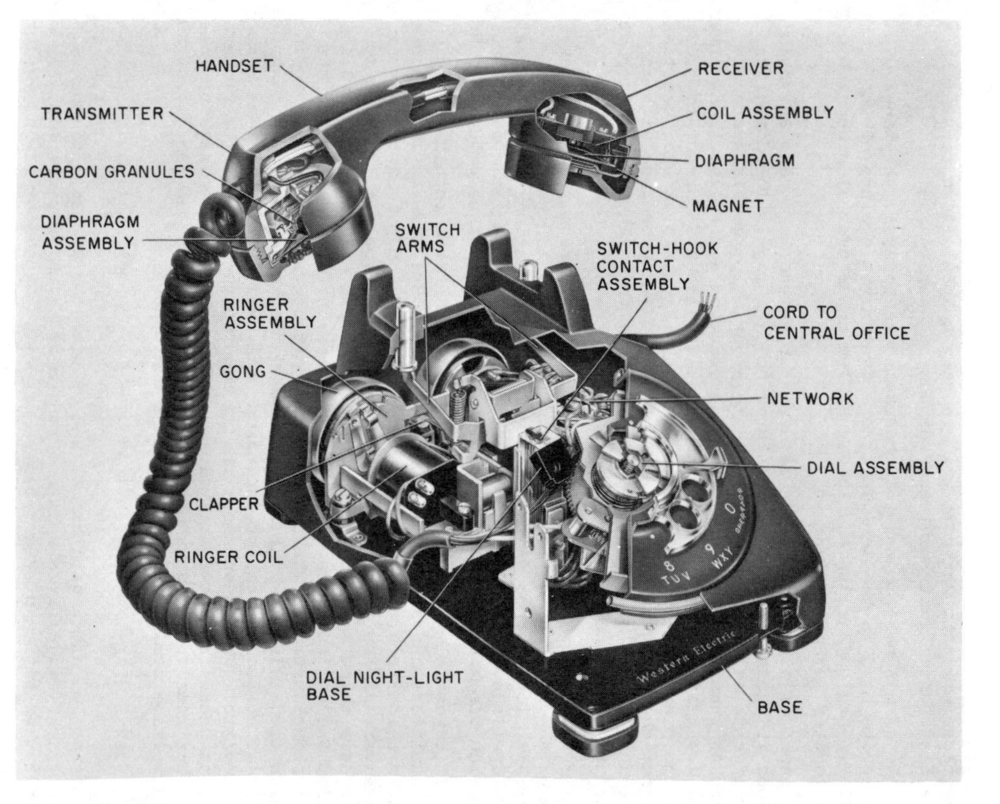

Figure 1-1 *The 500-type Telephone Used in the Bell Telephone System*

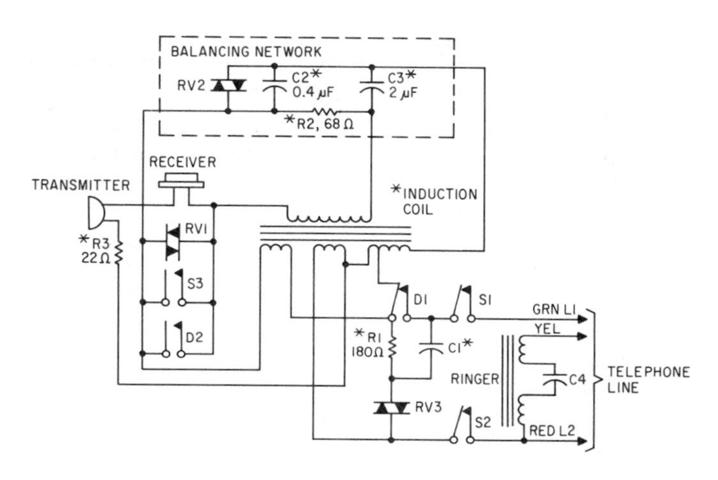

Figure 1-2 *500-type Telephone Schematic*

central office to maintain satisfactory transmission volume. All of the components marked with an asterisk (*) are located within the network block located inside Western Electric and ITT telephones. In phones produced by Automatic Electric, these components are contained on a printed circuit board in the phone.

While most telephones have three wires coming out of them — green (L1), red (L2) and yellow — very often the green and yellow wires are connected together. The yellow and red wires are used to make the telephone ring. To do this a voltage of between 60 and 90 volts, with a frequency of about 20 Hz is applied to the ringer, which consists of two coils in series with a capacitor. The ringing coils of the bell and capacitor C4 are designed so that once the bell starts to ring, the ringing signal is reinforced and even a small electrical current will cause the bell to ring. This is known as resonance. The ringer assembly is connected to the telephone line at all times because it is connected before the hook switch, which disables the rest of the telephone until the receiver is lifted off the hook. If the telephone company wants to determine how many phones are connected to a line, it either measures the capacity of the line or the current drawn during ringing.

When the receiver is lifted off the hook, switches S1 and S2 close, while switch S3 opens. The closing of switches S1 and S2 places the phone resistance, which may vary between 600 and 900 ohms, across the line. The 48 volts that are across the line when the phone is on the hook drop to about 5 volts.

To dial a phone number, the line must be interrupted (opened and closed) at a repetition rate of 10 pulses per second. This is done by contact D1 which is located on the back of the telephone dial. If, for example, the number 7 is dialed, D1, which is normally closed, will open and close 7 times. While all this is happening, D2 stays closed. The minute the dial is moved from its normal resting position, contact D2 closes and stays closed until the dial returns to its resting position. This short circuits the receiver to prevent the dial clicks from being heard.

2. Tariffs

There is a gray area of regulatory conditions surrounding the use of privately-owned telephone equipment on telephone company lines. The famous 1968 "Carterfone Decision," which permitted attachment of privately owned equipment to the telephone lines, was not a Federal Court decision as most people believe. The Federal Court referred the case to the Federal Communications Commission because of its technical nature. Thus the decision was only an FCC opinion, and not a court ruling.

That decision was against the restrictive practices of the telephone company and said, essentially, that the attachment of customer-owned equipment, which increases the utility of the phone to the user, should be permitted providing that such attachment does not pose a hazard to the telecommunications network, telephone company equipment, its employees, or the public.

As a result of the Carterfone Decision, telephone companies modified their tariffs, permitting such connections, but with wording inserted which requires a "voice connecting arrangement."

The FCC has since entered into hearings with the telephone companies, the state Public Utilities Commissions, and telephone equipment manufacturers to finally implement the intent of the 1968 decision. While moving slowly, progress has been made; and in most cases, the agreements reached and the tariff changes have all been in the direction of allowing private equipment to be connected to the line. AT&T itself has changed its tariffs recently to allow certain equipment, which contains a telephone company approved connecting device, to be connected to the line.

But these telephone company approved devices increase the cost of the equipment sold and many people feel they are unnecessary. Experts in the field find it illogical and unreasonable for AT&T to use foreign attachments without the "voice connecting arrangement," but to require it when the identical device is purchased directly by the telephone user.

An Electronics Industry Association staff vice-president for communications and industrial electronics has reported that more than 1,800 phone companies nationwide already use independently made attachments without the so-called protective devices for interconnect.

The FCC has indicated to AT&T that the Bell System should not require connecting arrangements for equipment when the same type of equipment provided by the telephone companies themselves is not connected through these voice connecting arrangements. The Bell System, for example, offers a telephone answering machine on a rental basis and does not connect that unit through the connecting arrangement they want to require others to use.

Many experts believe that it is now perfectly legal and permissible to connect telephone accesories to the regular phone line providing that these devices do not interfere with normal telephone operations, deprive the phone company of its lawful revenue, or create a hazard to its equipment or personnel.

The Bell System itself, in a manual entitled, *Voice Connecting Arrangements*, states in Section 2.21, "Responsibility of the Customer."

> The Tariffs permitting direct electrical connection of customer-provided terminal equipment state that:
>
> > Where long distance telecommunications service is available under this Tariff for use in connection with customer provided equipment, the operating characteristics of such equipment shall be such as not to interfere with any of the services offered by the Telephone Company, Such use is subject to to the further provision that the customer-provided equipment does not endanger the safety of Telephone Company employees or the public; damage, require the change in or alteration of the equipment of other facilities of the Telephone Company, interfere with the proper functioning of such equipment or facilities; impair the operation of the telecommunications system or otherwise injure the public in its use of the Telephone Company's service.
>
> . . . Upon notice from the Telephone Company that the customer-provided equipment is causing or is likely to cause such hazard or interference, the customers shall make such change as shall be necessary to remove or prevent such hazard or interference.

The FCC's position regarding interconnect is quite clear and has been expressed many times. Even where a potential for harm exists, the FCC is determined to investigate the facts and to assure that only minimum restriction and regulations will be allowed as are absolutely required to prevent such harm.

Many believe that it is only a matter of time before the restrictive interconnecting arrangements will no longer be required. For the time being, would-be users of telephone accessories have three choices:

1. Use only equipment that is inductively or acoustically connected to the telephone line.

2. Pay to have the telephone company install a connecting arrangement and pay a monthly service charge.

3. Ignore the tariffs and connect your telephone accessories directly to the telephone line.

3. Techniques

Some of the projects in this book have a printed circuit board layout associated with them. This is to make construction of the projects easier and to insure proper circuit hook-up. Those of you unfamiliar with printed circuit (pc) board fabrication may at first feel that the need for printed circuit boards would do just the opposite, i.e., make the project more difficult to build. However the pc board fabrication techniques described here should make you change your mind quickly. As a matter of fact, once you've made a few circuit boards, you'll probably wonder why you didn't do it before.

Hobby magazines have presented many articles on how to prepare pc boards, but all of them have had some drawbacks. This technique combines all the good points of these different methods into one.

To start a printed circuit (pc) board it is first necessary to have a pc board layout. That's no problem here because each project that uses a pc board comes with its own layout. The best thing to do is to first make one or two same-size photocopies of the layout, but you can work with the one in the book if you want to.

The layout should be cut out and pasted onto a piece of cardboard or heavy poster paper. Next, a hole puncher is used to make a hole at each point where there is a circular pad for a component to be mounted as in Figure 3-1. Once all the holes are punched, take the cardboard mounted pc layout and place it on a piece of copper-clad circuit board. Make sure that the copper side is up, so that copper can be seen through the holes in the pc layout.

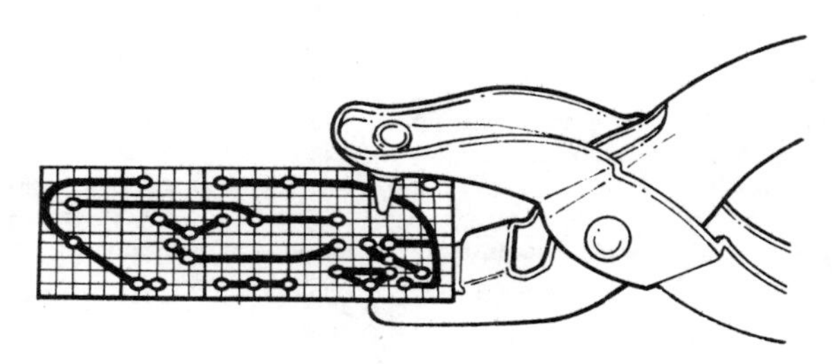

Figure 3-1 *Hand Punch Making Holes on a PC Board Layout*

If the board is too large, it should be cut down to size. With a felt resist pen (available from any hobby electronics supplier) make resist dots on the copper using the holes in the layout as a guide. When you are finished, check and see if there are any holes that show copper instead of the black resist. If there are, fill them in. Allow the resist to dry for a few minutes and then remove the layout mask.

There are now two ways you can complete the board. You can either make the dots a little bigger and interconnect them as in the pc layout with the resist pen, or you can get some direct etch pc patterns from Datak, also sold at electronic hobby stores, and place a precut pad on each dot and interconnect dots with the tape provided.

The first approach is less expensive and yields a perfectly workable board. The second approach, while costing more, results in a more professional looking board.

In either case, once the etchant resistant material is on the board it is ready for etching.

Etchant is available in two forms, liquid and powder. The liquid is convenient and ready to use, but the powder is a little better because it is stronger and when it is prepared, the chemical reaction that takes place when water is added to it increases the temperature of the resultant solution. This makes etching faster.

Pour the etching solution into a plastic tray and place the pc board into it. After a few minutes the copper becomes dull and a film will be built up on it. The copper that is exposed is dissolving. It is very important that you constantly agitate the tray in which the etching is taking place so that this film is continually removed from the surface of the copper. Agitating also causes more etchant to come in contact with the surface of the exposed copper and thus helps decrease etching time.

While the directions given with the various etchants available generally indicate that a board should be completely etched within 15 to 30 minutes, practical experience has shown that it can take as long as an hour. Of course a lot depends on the thickness of the copper being etched.

Throughout the etching process it is more important that contact with the solution be avoided. Plastic tongs should be used to handle the pc board if necessary. If some of the etchant does splash on you rinse it off right away with water. Try to wear old clothes while working with etchant, because its stains are very difficult to get out.

Once the etching has been completed and no more bare copper is visible, the board should be removed from the etching solution and rinsed off with water. Afterwards, take some steel wool and clean off the resist that is covering the copper areas you didn't want etched away.

Now take a very fine drill bit (this usually comes with a printed circuit board kit as does the etchant and copper board) and drill a hole at every point where there was a dot made through the layout mask. These holes will allow you to mount the circuit components on the opposite side of the board. Drill any mounting holes that may be required with the appropriate size drill. You are now ready to assemble your circuit.

A helpful way of assembling a pc board is to mount all of the passive components such as resistors and capacitors, first. Then mount the diodes, transistors, and finally, integrated circuits.

When working with integrated circuits it is best to use sockets. These will increase the cost of your project, but will facilitate the removal, if necessary, of the circuits.

Still, it is possible to remove these devices. The best way is to use braided wire like that found on shield cables, to suck up the solder. It is sold in little rolls in electronic hobby stores. If you can't get it, then just use some cable shielding.

When using this solder remover, put it over the solder joint to be freed and apply your soldering iron to it. You'll see the solder being sucked up as if this wire were a wick. When it becomes saturated with solder, just snip off the small piece involved and continue.

When inserting components into a pc board, be extra careful to follow component placement diagrams carefully. Observe polarities that appear on capacitors, diodes, transistors, and, most important of all, integrated circuits.

Never use an acid core solder. Always use a high grade of noncorrosive, 60/40 alloy, flux-core solder. Make sure it has a small diameter, such as a No. 20. Use only a small soldering iron with a rating of 40 watts or less. Do not use a transformer-type soldering gun. Always use extreme care when making solder connections to integrated circuit leads.

If you haven't had previous experience building electronic circuits, practice soldering on some pieces of scrap wire. Excess heat from a prolonged application of even the low wattage iron suggested, can permanently damage a solid-state device. Always try to connect some sort of heat sink, such as an alligator clip, to the leads of semiconductor devices. Also, always keep the tip of your soldering iron clean. A wet sponge can do this.

When soldering components to the copper foil of the pc board, place the soldering iron against both the wire and the foil. Apply the solder to the tip of the iron and to the junction point of the component lead and the copper foil. Use only enough heat to cause the solder to flow out and cover the circular section of the wire through which the wire protrudes.

Two projects in this book require the use of very sensitive CMOS integrated circuits. These require special handling. Since CMOS devices are very sensitive to even small static currents, they come specially packaged in a conductive package. This may take the form of a special plastic or aluminum foil. Therefore, CMOS devices should not be removed from their holders until they are ready to be used. Also, do not store them in plastic trays that are nonconductive.

When soldering such devices, always be certain to ground the tip of your iron. Also ground any test equipment that you may attach to those devices.

To illustrate some of the concepts explained and to provide you with a power supply that will come in handy for the projects in this book, let's go through the fabrication of a regulated power supply.

The schematic for the power supply is shown in Figure 3-2. The ac voltage is stepped down by the transformer to 24 volts ac and then rectified by the bridge rectifier. The rectified voltage is then fed to a resistor/zener diode circuit, where a reference voltage is produced, and to the collectors of Q1 and Q2. The zener reference voltage, which determines the output voltage of the supply, is applied to the base of Q2 which is connected in a Darlington configuration with Q1. Output is taken across the series combination of the 270 ohms resistor and the 0.1 μF capacitor.

The components for the supply are not critical. Just about any NPN power transistor can be used instead of the 2N3055 specified, and any low frequency small signal NPN can be used for Q2. The output voltage of the supply equals the zener voltage minus the voltage drop across the two transistors, or $V_z - 1.2$. To change the output voltage of the supply, simply change the zener diode.

If a large amount of current is required, the transistor will become very hot and it may be necessary to mount it on a heat sink. As long as you can hold a finger on the transistor for a minute while it is operating, you're okay. If you have to pull your finger away immediately, use a heat sink. But be careful, don't burn yourself.

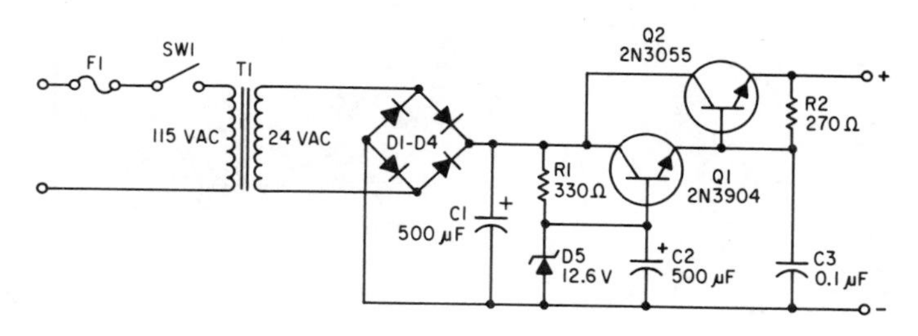

Figure 3-2 *Power Supply Schematic*

Parts List

R1 — 330 ohms
R2 — 270 ohms
C1 — 500 μF 50V electrolytic
C2 — 500 μF 16V electrolytic
C3 — 0.1 μF
D1-D4 — 1 amp rectifiers 50V PIV
D5 — 12.6V zener diode

Q1 — 2N3904 or equivalent npn
Q2 — 2N3055 or equivalent npn power transistor
T1 — power transformer, 115V primary, 24V secondary rated at 3A
F1 — 1 amp, 115V fuse
SW1 — spst toggle switch

4. Remote Ring Indicator

Ever miss a phone call because you were in the garage or bathroom and couldn't hear the telephone ring? Know anyone who is hard of hearing and could benefit from an extra loud bell or a flashing light that indicates the phone is ringing?

If you answered yes to any of these questions you have two choices. Either you can pay the telephone company to install an extension telephone or bell, or you can build one yourself. The do-it-yourself route has several distinct advantages. First you don't have to pay installation and monthly rental charges. Second, you're a lot more flexible. By connecting two wires to the contacts of relay RY1 (Fig. 4-1), you can place the extra bell anywhere you want and move it at will. Finally, if you decide that you want to build an automatic telephone answering machine, all you have to do is connect this circuit to others that are described later in the book.

About the Circuit

The remote ring indicator is basically a sound switch. When a call comes in, the crystal microphone of the indicator amplifies the ringing of the bell and allows a charge to build up across capacitor C1. When the voltage across C1 reaches the triggering threshold of the SCR, the SCR fires and closes relay RY1. The relay can then be used to control any other indicating devices. Since the triggering of the SCR produces a latching type action, any device connected to the contacts of RY1 will operate continuously. To reset the SCR and relay, simply ground the base of transistor Q1 by momentarily pressing the normally open pushbutton SW1. This causes Q1 to cut off, reducing the current flow to the SCR and causing it to unlatch.

For intermittent operation that can be synchronized to the ringing of the telephone bell, the additional circuitry in the dashed box is necessary. This circuitry is simply a unijunction transistor oscillator. Instead of manually resetting the SCR, a negative pulse from the oscillator is applied to the base of Q1 which resets it automatically.

The frequency of the oscillator, and hence the reset pulse, is determined by time constant $T = R1C2$ with $f = 1/T$. By adjusting R1, it is possible to have the relay and SCR reset after each ring.

Construction

The remote ring indicator can be constructed using a perforated phenolic board and mounted in an aluminum chassis that measures at least 5" × 9" × 2". This chassis, while much larger than needed to

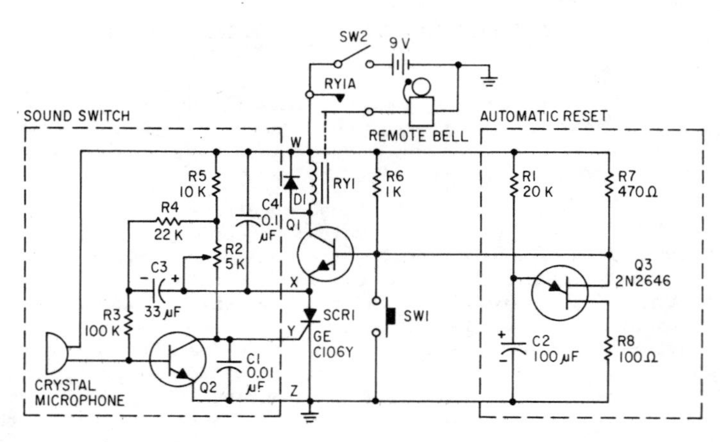

Figure 4-1 *Remote Ring Indicator Schematic*

house the printed circuit board and relay, is convenient because it allows the standard 500-type phone to sit on top of it. The chassis also has enough room in it to hold the additional circuit boards needed to build the automatic telephone answering machine.

When mounting the printed circuit board, the crystal microphone is placed about 2 inches from the rear of the chassis. This is where the sound pick up for the telephone bell is best, because that is where the bell of a telephone sitting on top of the chassis is located. Drill a hole about ⅛ inch in diameter in the center of the chassis on 2-inch line and mount the circuit board so that the microphone is directly underneath the hole. Next mount the relay, reset pushbutton SW1, ON/OFF switch SW2, and 9V battery.

Parts List

R1 — 20k ohms	C3 — 33 μF 16V electrolytic
R2 — 5k trimming potentiometer	C4 — 0.1 μF
R3 — 100k ohms	MIC — crystal microphone
R4 — 22k ohms	Q1 — 2N3904 or equivalent
R5 — 10k ohms	Q2 — 2N3904 or equivalent
R6 — 1k ohms	RY1 — 6V relay
R7 — 470 ohms	Q3 — 2N2646 unijunction transistor
R8 — 100 ohms	SCR — GE C106Y or equivalent
C1 — 0.01 μF	SW1 — spst momentary pushbutton
C2 — 100 μF 16V electrolytic	SW2 — spst toggle

Installation and Operation

To check the unit, turn on the power switch and tap the chassis lightly. You should hear the relay kick in. If it does not, adjust the threshold control R2 until it does.

If the relay kicks in as soon as the unit is turned on, press the reset button. If it still stays on, the threshold is not high enough and must be adjusted with R2.

Once the threshold is properly set, have someone telephone you. The indicator should kick in after the first ring; if it does not, the threshold isn't set properly. Once you have the proper setting do not worry about it going off accidentally. Unless someone hits the chassis, the remote ring indicator will not trigger falsely. The reason for this is that the chassis itself acts as an acoustical shield which prevents outside noises from setting off the device.

The remotely operated device can be connected directly to the extra terminals on RY1 or to another relay that is controlled by these terminals.

5. Teleswitch

How would you like to be able to call your house after an evening out, turn the electric coffee pot on, and have a fresh pot of hot coffee waiting for you when you get home? Or maybe you'd like to turn the lights on and off in your house while you are away on vacation so potential burglars won't realize that you are not home.

You can do these things and more with a teleswitch. If you want to, you can turn on a whole series of devices in sequence, just by making one phone call every time a device is to be turned on.

And the best part of the whole thing is that you do not get charged for a phone call, even if it is long distance. The reason for this is that the teleswitch does not cause the phone to be answered, it simply uses the ring signal to activate whatever devices are connected to it.

Do not worry about anyone else turning things on accidentally. The teleswitch is designed so that unless the phone rings exactly once, nothing will happen.

There are two versions of the teleswitch: sequential, multiple-device and on/off switching. The sequential type will turn on a series of electrically operated appliances one after the other. This is good if you have several things to control remotely such as an electric coffee pot, warming tray, lights, etc. The disadvantage of this device is that to turn something off, it requires a separate relay in the sequence chain.

This turn-off problem can be eliminated if you are only interested in turning one device or group of devices on and off together. This configuration uses a flip-flop circuit to control a relay that turns whatever is connected to it on and off. The first time you dial, it turns the device on. The next time, it turns it off.

About the Circuit

If you have read the book straight through from the beginning, by now you are familiar with the circuits that form the basis for the teleswitch. This device uses two sound switches, similar to the one described in the remote ring indicator in Chapter 4.

As was described earlier, when the telephone rings the crystal microphone picks up the sound of the bell and triggers SCR1 which closes relay RY1 (Fig. 5-1). Transistor Q1 is held on by resistor R1 and is used to reset the sound switch by applying a negative pulse to its base.

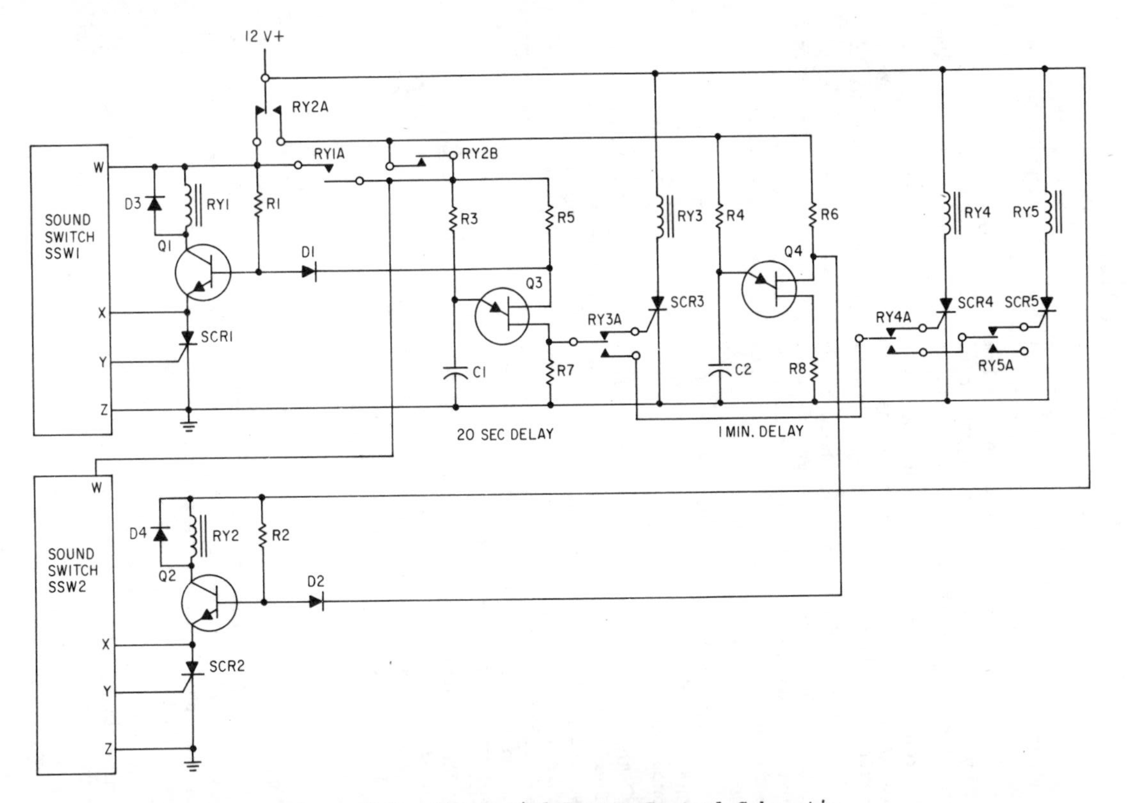

Figure 5-1 *Sequential Remote Control Schematic*

When RY1 closes, its contacts apply voltage to the second sound switch SSW2 and to a 20 second unijunction timer circuit.

If the telephone rings only once, which is what happens if you place a call to turn something on, capacitor C1 has enough time to charge up and trigger unijunction transistor Q3. The time required for C1 to charge up is determined by the R3C1 combination. The values shown will give a delay of about 20 seconds.

When Q3 is triggered, it produces a pulse that is used to turn SCR3 on and Q1 off. SCR3 latches relay RY3 on. One set of contacts on RY3 is used to prepare the circuit to trigger SCR4 on the next signal. This is done by transferring the gating pulse from SCR3 to the gate of SCR4, etc. The other set of contacts are used to control the first item to be turned on.

While all this is happening capacitor C2 is building up a charge. When the charge on C2 reaches the triggering voltage for unijunction Q4, a reset pulse is generated that resets SSW1 if the call was not a legitimate control signal. The components used produce a delay of about 1 minute from the time the phone first rings until the reset pulse is generated.

What this means is that sequential devices cannot be activated unless there is an interval of at least 1 minute. This delay prevents accidental activation by random phone calls.

Parts List

R1 — 1k ohms	C1,2 — 100 μF 16V electrolytic
R2 — 1k ohms	D1-4 — 1N914 or equivalent
R3 — 200k ohms	Q1,2 — 2N3904 or equivalent
R4 — 680k ohms	SCR1-6 — GE C106Y or equivalent
R5,6 — 470 ohms	RY1-6 — 12V dpdt relays
R7,8 — 100 ohms	Q3,4 — 2N2646

For on/off switching, the second version of the teleswitch, shown in Figure 5-2, should be used. This circuit is similar to the sequential type in that it uses two sound switches and two unijunction transistor timing circuits. The difference between the two is that the output of Q3 is used to trigger a flip-flop instead of an SCR. The negative going pulse from Q3's B2 terminal does two things. First it goes via isolating diode D2 to the base of Q1 and turns off SCR1 and its associated relay. The negative pulse is also applied via isolating diode D3 to the triggering circuit of the bistable flip-flop. C3 blocks any dc levels while diodes D5 and D6 serve as steering diodes which cause pulses to change the flip-flop from one state to another, forming the on/off switching action.

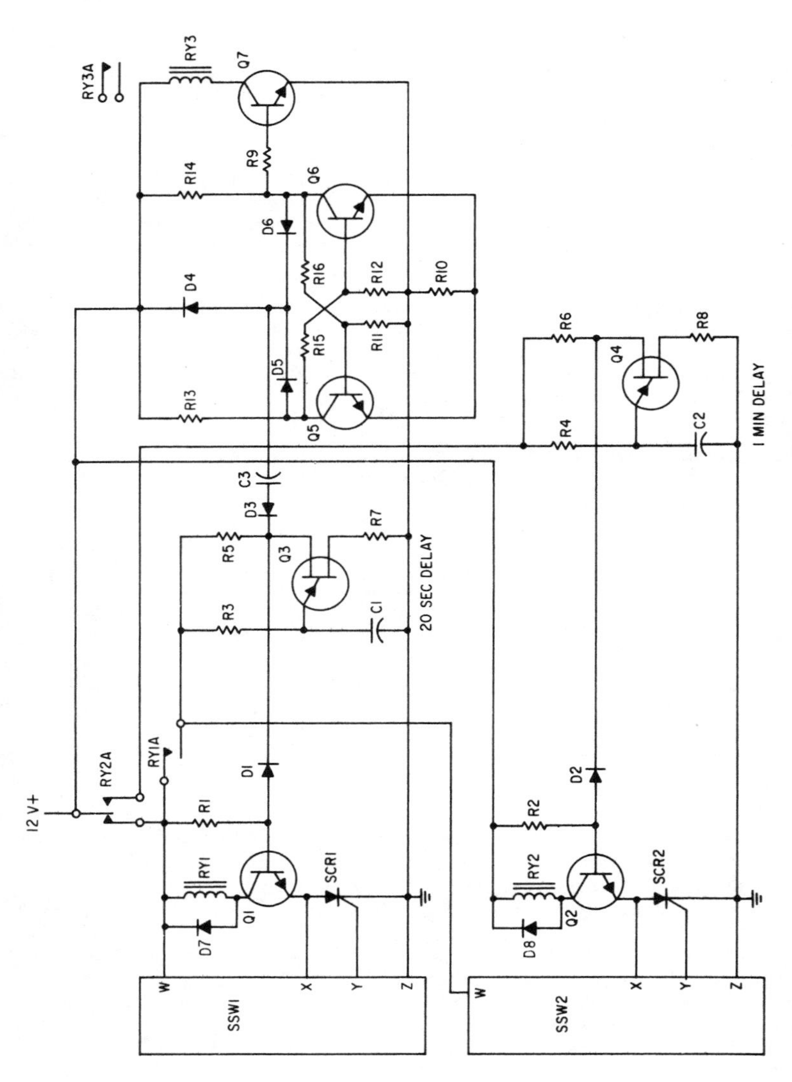

Figure 5-2 *On/Off Remote Control Schematic*

The output from one side of the flip-flop is fed to Q7, which acts as a relay driver and closes the relay on every other pulse.

As with the first version of the teleswitch, this one requires exactly one ring of the telephone to become activated. More than one ring will trigger SSW2 and disable the entire system for one minute.

Parts List

R1,2,9 — 1k ohms	C1,2 — 100 µF 16V electrolytic
R3 — 200k ohms	C3 — 1 µF 16V (electrolytic or non-electrolytic)
R4 — 680k ohms	D1-8 — 1N914
R5,6,10 — 470 ohms	Q1,2,5-7 — 2N3904
R7,8 — 100 ohms	Q3,4 — 2N2646
R11,12 — 10k ohms	SCR1,2 — GE C106Y
R13,14 — 3.9k ohms	RY1,3 — spst 12V relay
R15,16 — 33k ohms	RY2 — spdt 12V relay

Construction

The teleswitch, like many of the other telephone accessories described in this book, is best built in a chassis that measures at least 5" × 9" × 2". Depending on how many sequential devices you are going to control, you may want to use a larger box, with enough room to mount all of the controlled outlets.

Both sound switches should be mounted next to each other and located in the middle and towards the rear of the chassis.

Two 1/8-inch holes should be drilled at the spot where the crystal microphones will be mounted so that the ring signal can be picked up more easily. The controlled outlet(s) are mounted next, after you have first drilled a 1 5/16-inch hole to accommodate the outlet recommended. This outlet is similar to the ones commonly found in homes except that it does not have two receptacles, only one. It can be purchased in any electrical supply store. After drilling the main hole for the outlet, make the two small holes for the retaining screws. Remember to choose the chassis size according to the number of controlled outlets you are going to have.

Layout of components requires no special attention. Perforated phenolic circuit boards can be used, and any convenient arrangement will do. The 12V power supply described for Figure 3-2 should be used. If a large number of high-current relays are going to be used, make sure the supply can handle all the current required; otherwise a higher current transformer and power transistor will be needed.

Installation and Operation

To use the teleswitch it is only necessary to place the telephone on top of the chassis and plug in the devices to be controlled. As in other applications that use the sound switch, it is necessary for you to adjust the threshold level so that the switch triggers on the telephone ring. This is done by adjusting the value of the potentiometer on each sound switch so that the SCR will latch after the first ring.

If the sequential teleswitch is used and you want to turn things off remotely, remember that you have to connect that device to the power line through a normally closed relay so that the relay contacts can be opened when the relay is activated.

To test out the unit, have someone telephone you, but do not answer the phone. Let it ring a few times and then have the caller wait 1 minute and call back.

If the unit is operating properly, a relay click should be heard after the first ring and another relay click after the second ring. If so, the controlled device should remain off. After a total of 1 minute from the first ring, the sound of a relay opening should be heard. This is RY2 disconnecting power from the reset timer. Now the teleswitch is ready to accept a new ring signal.

Have your friend call again, this time allowing the telephone to ring only once. Twenty seconds after that ring, the device connected to the controlled outlet should turn on.

For the sequential version of the circuit, the next single ring should turn on the second outlet. In the on/off version, the next single ring of the telephone will turn the controlled outlet that is now on, to off.

6. Remote Ear

Did you go on a trip and forget whether or not you turned off the television or left the water running? Maybe you just want to check your house and see that everything is quiet and no one has broken into it.

If you want to do any of these things, you can with the remote ear. The remote ear is an adaption of the teleswitch circuit. It will automatically connect a microphone and amplifier to the telephone so you can monitor a remote location. As you will quickly see from Figure 6-1, the remote ear uses the same type of signal detectors as the teleswitch. However, instead of having controlled outlets to turn devices on and off, the remote ear has a small three-transistor amplifier connected to it.

The amplifier is identical to the one described in the speakerphone circuit, and since it uses a crystal microphone, it features a very clear and audible signal. The output of the amplifier, which is located in the area that you want to monitor, is fed to a small speaker which is acoustically coupled to the telephone mouthpiece.

Since it is unlikely that remote listening will be done for long periods of time, the remote ear has a built-in timer that allows you to listen for a specified amount of time, about 3 minutes. Longer or shorter times can be achieved by adjusting the timing resistor R6 of Q4. After the selected time period, the remote ear automatically hangs up the telephone.

About the Circuit

Sound switch 1 (SSW1) and the unijunction timer that is associated with it (Q3) are the same as were used in the teleswitch. Sound switch 2 (SSW2) however, is slightly modified. Instead of having one relay connected to SCR2, there are two relays, RY2 and RY4.

The operation of the remote ear involves several steps. When the telephone rings the first time, SSW1 triggers and causes RY1 to close. This applies power to the two timing circuits made up of Q3 and Q4.

If the phone rings more than once, SSW2 triggers and RY2 closes, disconnecting power from the first unijunction transistor timing circuit and from the first sound switch. This prevents the remote ear from being activated and makes it necessary to wait 3 minutes before the next attempt is tried.

If, however, the phone rings only once, there is enough time for charge to build up on C1 and for Q3 to trigger, turning RY3 on.

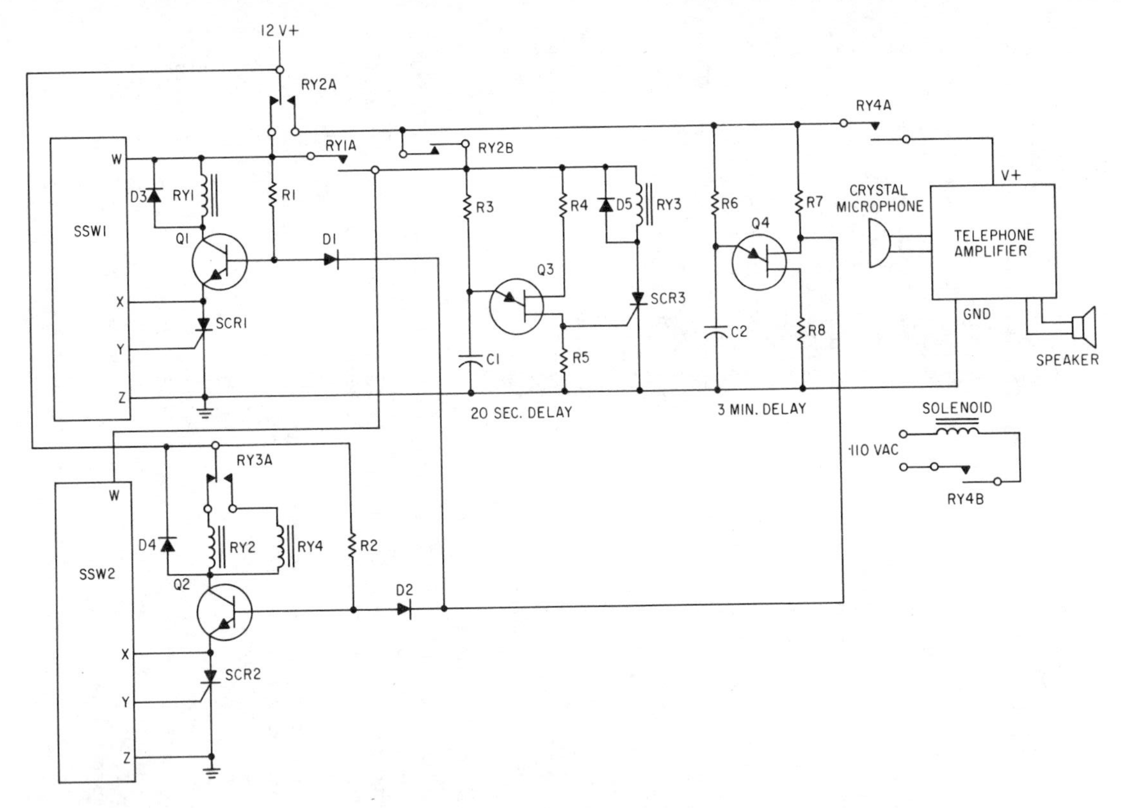

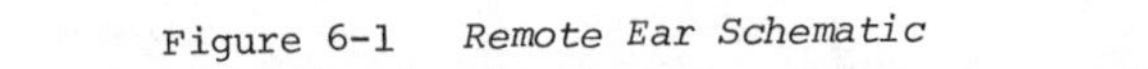
Figure 6-1 *Remote Ear Schematic*

When RY3 is activated it switches RY2 out of the control circuit of SSW2 and replaces it with RY4.

Since the first time the telephone rings only once it arms the circuit, the next time the telephone rings it will turn on the listening circuitry. This is done by SSW2 turning on RY4, which in turn controls the amplifier and the answering solenoid.

RY4 latches closed and is held in that position until a reset pulse from the unijunction timer Q4 turns transistor Q2 off and unlatches SCR2.

The actual answering of the telephone is done by a solenoid that pulls up when RY4 closes. This releases the cradle switch and answers the phone. The handset of the telephone is placed on the table along side the telephone. The loudspeaker that is connected to the output of the amplifier is held next to the mouthpiece (rubber bands can be used). Thus the sound picked up by the crystal microphone is amplified and acoustically coupled to the telephone.

After 3 minutes, or whatever time period you selected for the unijunction timer has elapsed, a reset pulse is generated by Q4 and the bases of Q1 and Q2 are brought to ground potential, turning these transistors off and unlatching the SCRs. The unit is now ready for its next monitoring period.

Parts List

R1,2 — 1k ohms	SCR1 — GE 106Y
R3 — 200k ohms	SCR2 — GE 106Y
R4 — 470 ohms	SCR3 — GE 106Y
R5 — 100 ohms	MIC — crystal microphone
R6 — 1M ohms	RY1 — 12V spst relay
R7 — 470 ohms	RY2 — 12V dpdt relay
R8 — 100 ohms	RY3 — 12V spdt relay
C1 — 100 μF 16V electrolytic	RY4 — 12V dpst relay
C2 — 200 μF 16V electrolytic	SPRK — 8 ohm speaker
D1-D5 — 1N914	SSW1 — sound switch from remote ring project
Q1 — 2N3904	SSW2 — sound switch from remote ring project
Q2 — 2N3904	SOL — 110VAC solenoid
Q3 — 2N2646	
Q4 — 2N2624	

Construction

This project is constructed from four modular circuits. The first two are sound switches identical to those built in an earlier project in this book (see Chap. 4). After the sound switches are built, they should be mounted in a metal chassis that is large enough to be placed under the telephone. A 5" × 9" × 2" aluminum chassis was used for the prototype. A 1/8-inch hole should be drilled where each of the crystal microphones are mounted so that sound will reach the microphones more easily.

After the sound switch modules SSW1 and SSW2 are mounted, put together the control module using the circuit shown in Figure 6-1. The circuit can be fabricated by wiring the components together on perforated board or a printed circuit pattern can be made.

After the control module is completed, mount the board, using spacers, at any convenient spot under the chassis. The only part missing now is the amplifier. The amplifier used here is identical to the one used for the speakerphone. This module too should be mounted to the chassis at a convenient location with spacers.

Once all four modules are mounted, mount the relays. Now connect all of the wires from the modules and the relays that go together directly to the positive side of the supply. If an external battery is going to be used to supply power, connect these leads to a screw terminal that is insulated from the chassis. If the power supply described at the beginning of the book is used, connect the leads to the positive terminal of the supply. Do the same for all ground leads. Connect all remaining wires to their proper locations.

Cut a piece of 1" × 2" wood to a length of 10 inches. This will be used as a vertical support for the solenoid which will hold the phone in the unanswered position until the proper command signal is given. To position the arm, place the telephone on top of the chassis and move the wooden arm until it is parallel to the telephone cradle where the handset is normally placed. Mark the spot, because that is where you want to mount the arm. Do so with at least two screws. Next place the solenoid on the inside of the arm and take the handset off of the telephone. Position the solenoid until the fully extended plunger holds the cradle switch down. Attach the solenoid to the wooden support with two screws.

Now attach the two conductors of conventional lamp cord to the two terminals on the solenoid. Then cover these terminals with electrical tape to eliminate touching them and getting a shock. Bring the lamp cord down the support, attaching it to the wood in several places with staples. Be careful that a staple does not pierce the insulation of the wire, causing a short circuit. Bring the wire into the bottom portion of the chassis through a grommet-lined hole and attach one of the two strands to one set of normally open contacts on RY4. Attach another piece of single conductor lamp cord to the other contact of the set. This wire, along with the unused wire from the solenoid, will be connected to the ac line.

Mount two miniature jacks to the chassis for the microphone and the speaker. The speaker can be acoustically coupled to the telephone by simply holding it next to the telephone mouthpiece with a few rubber bands. The microphone should be located at the spot you wish to monitor.

Installation and Operation

Installing the remote ear simply requires that you place the telephone on the chassis, remove the handset, and allow the solenoid to hold the cradle switch down. Now place the microphone in the room you want to monitor and turn on a radio. Place the speaker next to the mouthpiece of the telephone.

Have a friend call and ring the telephone several times. On the first ring RY1 should close. On the second ring RY2 should close, opening RY1. After 3 minutes, a reset pulse from Q4 should reset SCR2 opening RY2.

After another 3 minutes have passed and RY2 has reset, have your friend call again. This time tell him to ring the phone only once, and then call back 20 seconds later. On his second call, the phone should be answered automatically after the first ring. This is accomplished by RY4 which becomes activated by SSW2 when the phone rings the second time. RY4 closes the circuit to the solenoid and causes it to lift up, releasing the cradle switch of the telephone.

RY4 was activated because after 20 seconds elapsed, Q3 produced a pulse which triggered SCR3 and switched the power line from RY2 to RY4.

When the call is answered your friend should hear the radio playing. If he does not, check to make sure that the speaker is properly placed next to the mouthpiece of the telephone. Three minutes after the first ring, Q4 will generate a reset pulse and release RY4. This causes the solenoid to drop and hang the phone up. At the same time it opens the circuit to the amplifier. The remote ear is now ready to use again.

7. Speakerphone

No doubt you've had occasions when you placed a call to, or received a call from, someone you haven't spoken to for a long time, and other members of the family want to speak also. Or you were busy doing something and got annoyed because you had to drop everything and talk to someone on the phone for half an hour.

A speakerphone can be useful in both these situations. No longer need members of your family argue to talk first on the phone. Now everyone can set around the telephone and talk together with whomever is on the other end. The speakerphone will also allow you to carry on a conversation while you're still busy moving furniture or painting the walls.

And like most of the other projects in this book, the speakerphone requires no direct electrical connection to the telephone line. All coupling is made either by acoustic or magnetic means.

About the Circuit

The heart of the speakerphone is a three-transistor amplifier that takes the millivolt level signals produced by the telephone pickup coil and amplifies them to room-filling volume. The schematic is shown in Figure 7-1.

Just about any telephone pickup coil can be used for the speakerphone, but one of the newer ring-shaped devices is recommended because this particular coil gives a stronger signal and will also come in handy for some of the telephone accessory projects to

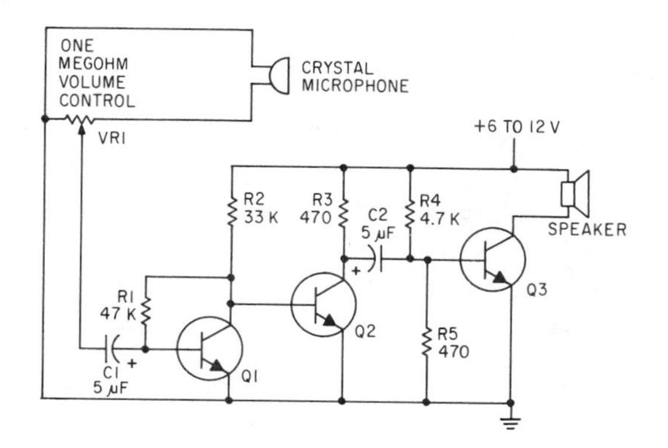

Figure 7-1 *Speakerphone Schematic*

follow. The coil is designed so that it will fit snugly over the receiver portion of the telephone handset, where it will pick up the audio signals to be amplified by induction.

The induced signals are applied to volume control VR1 where a fraction — depending on how loud you want the final signal to be — is tapped off the applied to the voltage amplifier made up of Q1 and Q2. After the voltage of the induced audio signal is raised by these two transistors, the signal is applied to Q3 where the current, and hence the total power level of the signal, is increased. Power transistor Q3 is capable of handling large currents. The need for an output transformer is eliminated by Q3 because its large current-handling capability means that it can interface directly with an output speaker.

So far we have talked about picking up a signal and amplifying it so that everyone in the room can hear it. That constitutes only half of the speakerphone. The other half of the speakerphone requires that people be able to talk into the phone from a remote location. This can be done in two ways. The first is by using another amplifier to pick up the sound generated in a room via a microphone. The second is to acoustically design the speakerphone so that sound focused in the direction of the device will be concentrated and fed into the mouthpiece of the telephone.

The first approach is most desirable but leads to many problems, the biggest of which is feedback oscillation. This of course could be eliminated by using voice-controlled switches, but that would mean conversations would have to be carried on in an artificial manner where one person would have to wait until the other stopped talking before he could be heard.

Simultaneous conversation with an amplified input is possible, but it requires complex circuit design and would considerably raise the cost and difficulty of construction.

To keep things simple, we'll do what most of the manufacturers of speakerphones do, and design the unit so that sound focused in the direction of the speakerphone is concentrated and directed towards the microphone of the telephone. This is not as difficult as it may sound. It is really quite simple and involves constructing the enclosure for the speaker so that sound is deflected upwards.

Parts List

R1 — 47k ohms
R2 — 33k ohms
R3 — 470 ohms
R4 — 4.7k ohms
R5 — 470 ohms
C1,C2 — 5 μF 15V electrolytic
Q1,2 — 2N3904 NPN
Q3 — 2N3055 NPN power transistor
VR1 — 1 megohm potentiometer
MISC — telephone pickup coil and hardware to construct box for project
SPKR — 8 ohm speaker

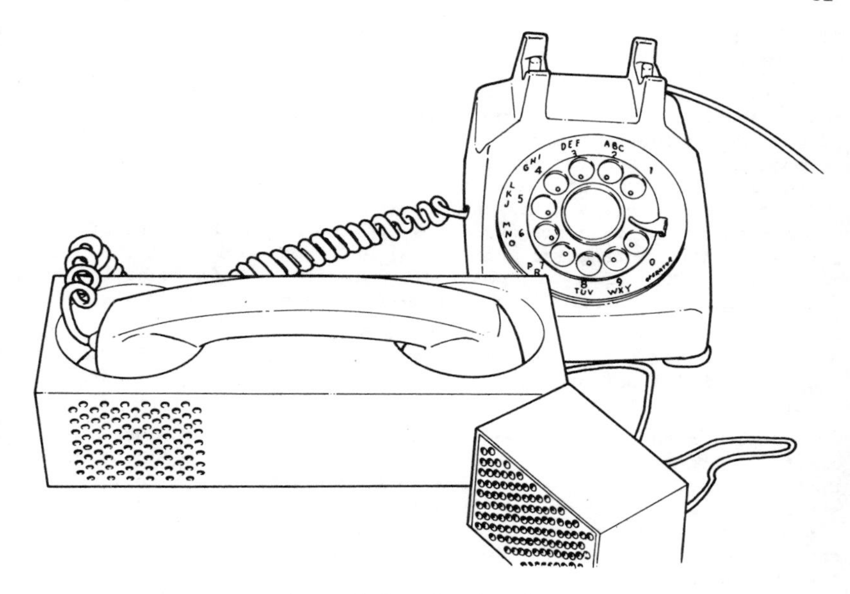

Figure 7-2 *Speakerphone*

Construction

In contrast to other projects described in this book, the enclosure is very important, for it is more than a case to hold the electronic circuit. It is a functional part of the design of the device (see Fig. 7-2).

To start with, build a thin, plywood box 10 inches wide, 4 inches deep, and $2\frac{1}{2}$ inches high. The top should have two holes cut in it $3\frac{1}{4}$ inches in diameter and made on center lines that are located 2 inches from each end.

These holes will be used to allow the handset to drop into the speakerphone and properly position it. With the box assembled and these two holes cut, mark one hole, either one, "microphone" and the other one "receiver." This can be done lightly in pencil and is only necessary to make instructions easier. On the front vertical wall of the box on the microphone side, draw a rectangle that measures $3\frac{1}{4}" \times 2"$ and place the rectangle so that it is $\frac{1}{4}$ inch from each of the three edges. Once marked, use a saber saw to cut out that rectangle.

This is the opening that will be used to direct sound into the microphone of the telephone. If you want, you can put a piece of grill cloth over the opening to improve its appearance, but make sure it is thin so that it will not severely attenuate the voice signals entering.

Next, 3½ inches from the inside of the microphone end of the box, place a barrier made from a piece of wood 4" × 2½". This will prevent sound entering the microphone opening from being dispersed throughout the box. To further concentrate the sound, take a thin piece of flexible plastic and cut it into a rectangle 3¼" × 5". This is the deflecting plate and should be inserted in the microphone compartment on a diagonal whose bottom is in the front of the box (where the rectangular opening is) and whose top is at the back of the box. The deflecting plate will need to be bent to be inserted. Bend it so the curvature of the plate is toward the bottom of the box.

Now, any sound entering the microphone compartment will be deflected upward toward the mouthpiece of the telephone. If done properly, you'll be surprised at just how effective this acoustic coupling can really be.

Proceed to the receiver end of the box. Here work is less critical. Mount a piece of thin plywood, cut into a 3¼-inch square, directly underneath the receiver hole so that when the receiver is placed in it, the surface of the wood is parallel to the receiver of the handset. The square can be held in position by fastening it to spacers, dowels, or wooden blocks. Next, mount the pickup coil on the top of the wooden square so that when the receiver is placed in the box, it will come into contact with the coil.

That's the hard part. The easy part is to fabricate the printed circuit — or use perf board if you wish — and construct the three-transistor amplifier. Mount the volume control at a convenient spot on the front panel along with an on/off switch. Mount the pc board, on spacers, at another convenient location. At a third convenient location mount a battery holder for the six penlight batteries.

The output of the amplifier should be fed to a miniature jack located on the side of the case. The speaker is in a little enclosure of its own and connected to the amplifier via a shielded cable and a miniature plug.

Installation and Operation

To install the speakerphone, it is simply necessary to place the handset of the telephone into the device, making sure that the microphone and receiver portions of the handset are inserted into the proper holes. Turn the unit on. You should immediately hear the dial tone coming through the speaker. If you don't, check the volume control and the coupling between the induction coil and the receiver.

Phone a friend and talk to him using the speakerphone. Try standing at various distances and at various angles from the microphone opening. Ask him to tell you which positions are best.

8. Scrambler

At one time or another most of us have the need or desire to pass along information that we would just as soon not put in writing or otherwise make public. To do so, we use the ordinary telephone; but many phones have extensions or some other means by which a conversation can be overheard. So keeping something totally confidential can be difficult.

If you really want to keep a phone conversation private, it is necessary to "scramble" your speech so that only the person for whom it is intended can understand it. The Security-1 does just that. When two parties are using this scrambler system and talking in plain language they can understand each other. But a third party listening in on an extension phone will hear a strange concoction of sounds that make no sense at all. It is impossible to decipher the conversation unless you have another scrambler and know the electronic key being used.

The Security-1 requires no electrical connections to the telephone — all coupling between the scrambler and the telephone is made by magnetic induction and acoustic means.

Besides the scrambler devices, the user on each end must have conventional audio sine-wave generators capable of delivering about 1 volte, tunable between 1 and 3 kHz. These are used as the scrambler sources. If a scrambling scheme that is almost impossible to decode is desired, the audio output from a conventional transistor radio, through the headphone connector, may be used as a scrambler source. In this case of course, both parties must be able to tune their receivers to the same broadcasting station.

About the Circuit

The basic principle of the Security-1 (Fig. 8-1) employs what is known as a balanced ring modulator. The same circuit is used for both coding and decoding. Each end of a scrambler system requires two telephone hand sets: the concentional house telephone (called the "house phone" here) and another handset (called the "project phone"). The project phone can be any surplus handset that has a conventional carbon microphone and dynamic earphone with a connecting cable.

With no speech applied to the primary of T1, when the applied encoding carrier is positive going (with respect to ground), the currents in the primary of T2 and the secondary of T1 (through diodes D1 and D4) are out of phase so that no carrier is developed in the secondary of T2. When the encoding carrier is negative going, the same thing happens as the current flows through output transformer T2.

When speech is applied to the primary of T1, the audio voltage across the secondary of T1 unbalances the diode modulator. The resulting signal across the secondary of T2 consists of a series of pulses whose polarity and repetition rate are determined by the carrier voltage and whose amplitude is determined by the instantaneous amplitude of the speech signal.

If the encoding carrier is assumed to be a 3000-Hz tone and the speech frequency is assumed to be a 100 Hz tone, then the output would contain both the 3100-Hz upper sideband and the 2900-Hz lower sideband. If a filter is used to cut off frequencies about 3000 Hz, then only the lower sideband remains. When the input speech frequency is changed to 200 Hz, the output will be 2800 Hz. Thus the modulator inverts the incoming speech frequency, making it completely unintelligible to the unwanted listener.

Decoding uses the same circuit as encoding and the system works as long as the same carrier signal is used at both ends.

Construction

The mechanical construction of the scrambler involves making a mounting for the house phone so that a pickup coil and a small loudspeaker can be placed in close proximity to the earpiece and the microphone, respectively, of the house phone. It is best to prepare the mounting first and then construct the electronic portion of the scrambler and fit it into the support.

The prototype uses a commercially available plastic telephone amplifier for the cabinet. You can build any type of cabinet (preferably made of wood) slightly longer than the telephone handset and a few inches deep. If you build your own cabinet, lay the house phone handset down on the upper surface and make the locations of the microphone and the earpiece. Cut out holes of the correct size so that the phone drops smoothly into place when it is in position.

Using appropriate hardware and spacers, mount the small 45-ohm loudspeaker under the microphone hole so that it is about one-half inch from the house phone microphone when the phone is placed in the support. Mount the induction pickup in the usual fashion to the earpiece. Any of the low-cost telephone pickup induction coils, available at most electronic supply stores, can be used here.

If you decide to use the commercial telephone-amplifier set, you will find all of these holes already made. You will also find an induction coil built into the earpiece hole. Remove the bottom cover of the cabinet, and take out the plastic insert from the microphone chamber. Then remove the built-in audio amplifier. Do not remove the induction coil. Also remove the small loudspeaker from its plastic cabinet. Using appropriate hardware and spacers, mount the loudspeaker in the microphone chamber as previously described. Although a 45-ohm speaker is specified in the parts list, you can use the lower impedance 8-ohm speaker that comes with the

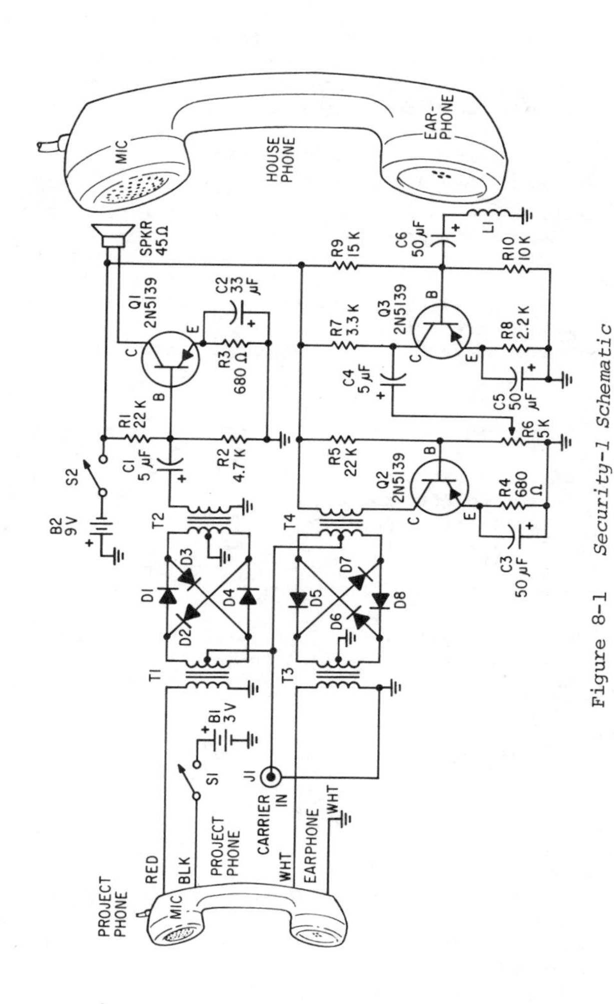

Figure 8-1 *Security-1 Schematic*

built-in amplifier. In this case, also remove the speaker output transformer from the amplifier printed circuit board and wire it to the speaker, using a pair of leads to run the primary back to the circuit.

In both the commercial and homemade cabinets, once the speaker has been mounted, use foam rubber to pad the perimeter of the microphone hole so that the house phone fits snugly in place. You can also insert foam rubber sound-deadening material under the speaker to keep the acoustic energy within the microphone chamber. In the commercial unit, leave the earphone jack in place; in the homemade unit mount an earphone jack on one wall.

If you are using the commercial telephone amplifier, most of the required components can be removed from the built-in amplifier including the transistors, volume control, and on/off switch, to be used in the scrambler. The driver transformer for the push-pull output stage can also be salvaged and used as T2. If you do not choose to use the pc board (Figs. 8-2 and 8-3), perf-board construction may be used, making sure that the overall board will fit within the enclosure.

Parts List

B1 — C or D cell (2)
B2 — 9V transistor radio battery
C1,C4 — 5 μF, 15V electrolytic capacitor
C2 — 33 μF, 10V electrolytic capacitor
C3,C5,C6 — 50 μF, 15V electrolytic capacitor
D1-D8 — small signal silicon diode (1N34A or similar or use RCA CA3019 IC)
J1 — earphone jack
L1 — telephone induction coil pickup (Lafayette 99E10340 or similar)
Q1-Q3 — small signal pup transistor (2N5139 or similar)
R1,R5 — 22,000 ohm
R2 — 4700 ohm
R3,R4 — 680 ohm
R7 — 3300 ohm
R8 — 2200 ohm
R9 — 15,000 ohm
R10 — 10,000 ohm
(All resistors ¼ watt)
R6 — 5000 ohm PC potentiometer
S1,S2 — spst switch
T1-T4 — 500 ohm to 500 ohm center-tapped transformer (TRIAD T34-X)
Misc. — telephone amplifier, surplus telephone, battery holders, transistor radio earphone cable and connector, audio signal generator, radio, mounting hardware, etc.

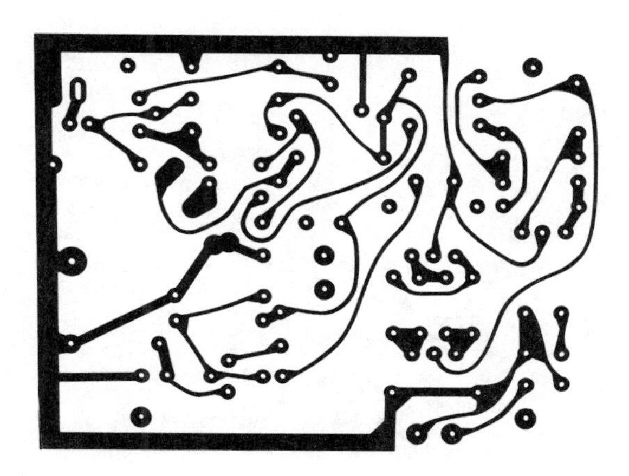

Figure 8-2 *Scrambler pc Layout*

The completed board is mounted on standoffs within the cabinet and a hole is drilled in the side of the enclosure that is large enough to pass the four-conductor cable from the project phone. In most phones, the two white leads are from the earpiece, while the black and red leads are from the microphone. Mount the batteries where convenient.

Installation and Operation

The scrambler can be tested without using the house phone. Disconnect both leads supplying the project phone mike to input transformer T1. Connect the loudspeaker output from any radio to the input terminals of T1 and tune the radio to an all news station, or one that has more speech than music. If you use a conventional radio, disconnect the speaker connections to the output transformer secondary and use the secondary to supply T1. If you are using a transistor radio, use the earphone hack that is usually provided. Turn the radio volume down.

Connect a conventional audio sine-wave generator, via a miniature phone plug, to the coder input jack on the scrambler. Set the audio generator to about 1 kHz, 1 volt. Turn on the scrambler power switch, S1. Slowly turn up the radio volume. Garbled speech will be heard from the built-in speaker.

By adjusting the radio volume control or signal generator output level control, the garbled speech can be heard at its "best" quality. If you adjust the signal generator frequency to about 3 kHz, the garbled speech will change. The best scrambling for the human voice takes place at about 1 kHz.

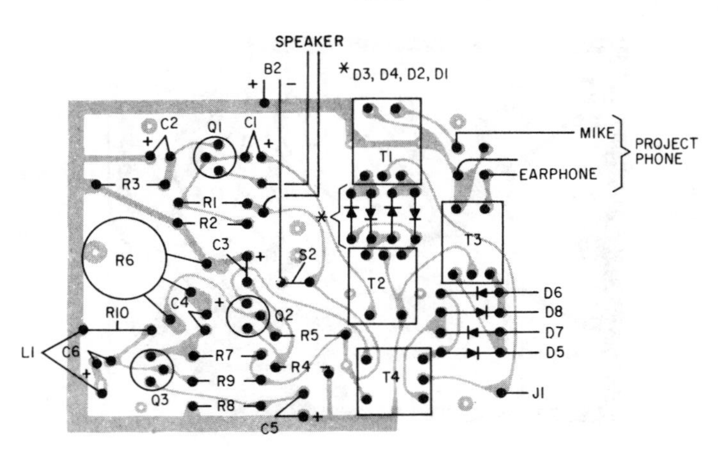

Figure 8-3 *Component Layout*

To test the unscrambler, connect the radio to the project phone earpiece leads and a transistor radio earpiece to the secondary of T3. When the project phone is placed in its correct position with the earpiece in the proximity of L1, scrambled speech will be heard in the radio earpiece. If audio tone breakthrough is encountered, connect a capacitor across the speaker to reduce the level of this unwanted signal.

When using the scrambler, dial as usual and tell the party on the other end that you are going to scramble. Make sure you have pre-arranged with him the audio frequency to be used or the radio station for scrambling. Then place house phones in the scrambler and talk through the project phones.

---

9. Digit Monitor

A very convenient device to have when you're dialing phone numbers, especially long ones, is a dialed-digit monitor. If you have ever started dialing a number and then been distracted for even a second, chances are you've had to hang up and start over again because you didn't remember what number you dialed last. That won't happen again if you build the dialed-digit monitor.

This project is different from the others in that it uses a number of integrated circuits instead of discrete transistors. Integrated circuits are a combination of many resistors, transistors, and diodes in one single package. It is the availability of these devices that makes this project possible.

In addition to integrated circuits, this project requires the use of a single-digit, light-emitting diode (LED) display. Both the integrated circuits and the LED should be available from local electronic parts suppliers. If not, they can be purchased by mail from any of the companies listed in the back of electronics hobby magazines.

About the Circuit

The dialed-digit monitor can be separated into two distinct parts: The pulse pickup circuitry; and the pulse counter and display (see Fig. 9-1).

Parts List

R1 — 5k ohms	RO1 — Man-4 seven-segment readout or equivalent
R2 — 3.3 Mohms	TC1 — telephone pickup coil
R3 — 680 ohms	C1 — 50 pF
R4 — 15k ohms	C2 — 0.01 μF
R5 — 5k ohms	C3 — 100 μF 16V dc electrolytic
R6 — 270 ohms	C4 — 0.01 μF
R7-13 — 470 ohms	C5 — 50 μF 16V dc electrolytic
IC1 — RCA CA3130 op amp	C6 — 50 pF
IC2 — 74121 one shot multivibrator	C7 — 0.01 μF
IC3 — 7490 decade counter	D1 — IN914
IC4 — 7475 latch	
IC5 — 7448 seven segment decoder	
IC6 — 74123 dual one shot multivibrator	

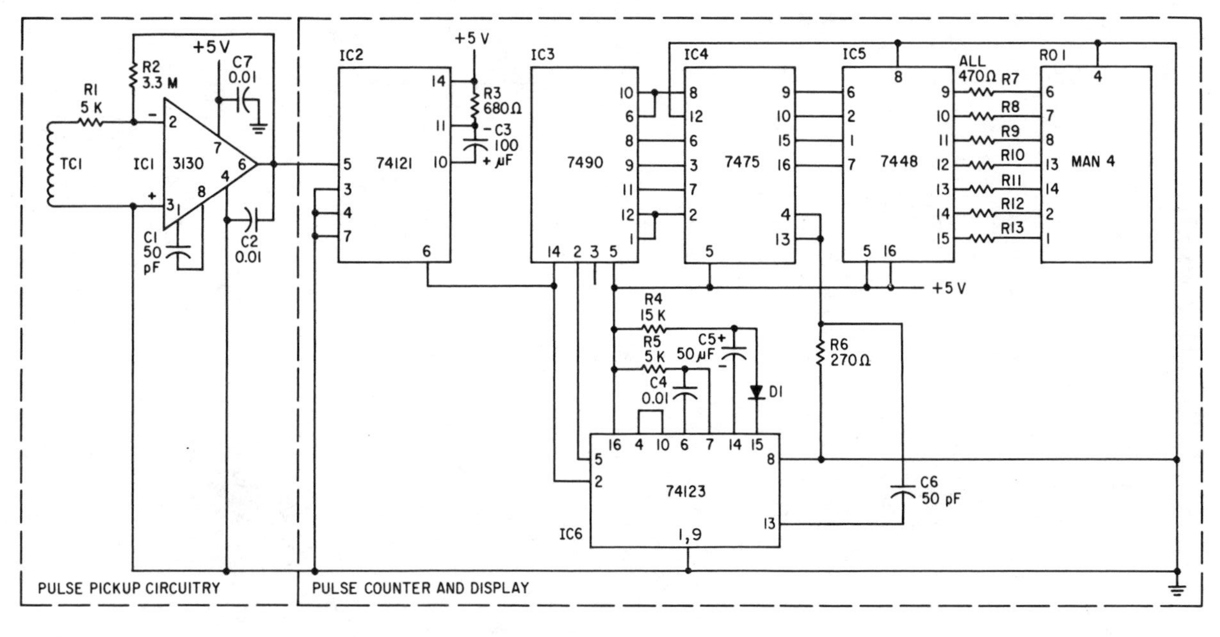

Figure 9-1 *Dialed-digit Monitor Schematic*

To pick up the dial pulses, we use a telephone pickup coil. This time, however, instead of placing the coil near the receiver portion of the handset, as is customary, it is placed underneath the telephone toward the rear right. When the telephone dial is used to dial a number, the dial generates a series of pulses, depending on the number dialed, that are used to open and close a switch that is in series with one of the two main telephone wires. Like voice signals in a phone, the pulses generate a magnetic field which can be picked up by the induction coil if it is properly placed.

The pulse signals, which are very weak, are fed to a very high-gain, integrated-circuit amplifier (IC1) that amplifies them 1,000 times. The gain of the amplifier is determined by the ratio of R2/R1.

Once the dial pulses are amplified they are fed to an integrated circuit known as a one-shot (or monostable) multivibrator (IC2). What this integrated circuit does is clean up the pulse and get rid of any noise associated with the original pulse. The width of the pulse is determined by the R3C3 combination. In this case we want a pulse that is less than 0.1 second long. So we set R3 to 680 ohms. The reason is the pulses generated by the telephone dial occur at a rate of 10 pulses per second. The values chosen will provide a pulse width of about 0.07 second.

The dial pulses are then fed to two more integrated circuits, IC3, which is a decimal counter, and at the same time to half of IC6, which is a dual one-shot multivibrator.

The decimal counter will count up to ten electrical impulses and then transfer the binary-coded information to IC4, which is a latch that remembers the number stored in it until it is reset. The information is transferred from the counter to the latch when the first one-shot multivibrator determines that it has not seen a dial pulse for at least 0.25 second.

As long as pulses are coming from the dial at 10 pulses per second, the one-shot stays off and nothing happens because the time between pulses is less than 0.25 second. Once dialing of a particular digit is over, more than 0.25 second elapses before the next digit can be dialed. At this point, the one-shot knows that you have finished dialing one digit and are about to start the next. So it generates a pulse that goes to the latch and tells the latch to remember the number from the decimal counter. Fifteen microseconds after the latch remembers the information at the output of the counter, a pulse from the other half of IC6 resets the counter to zero and it is ready to count the pulses from the next number being dialed.

In the meantime, the latch remembers the last number and sends that information to IC5, which converts the binary information into a code that lights up the proper bar segments on the light emitting-diode display.

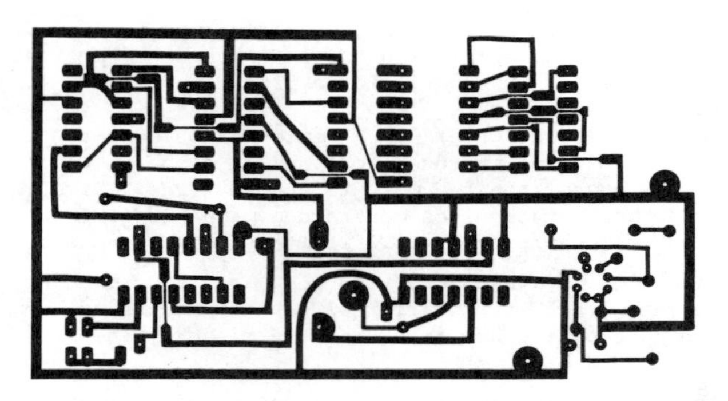

Figure 9-2 *Pc Layout for Dialed-digit Monitor*

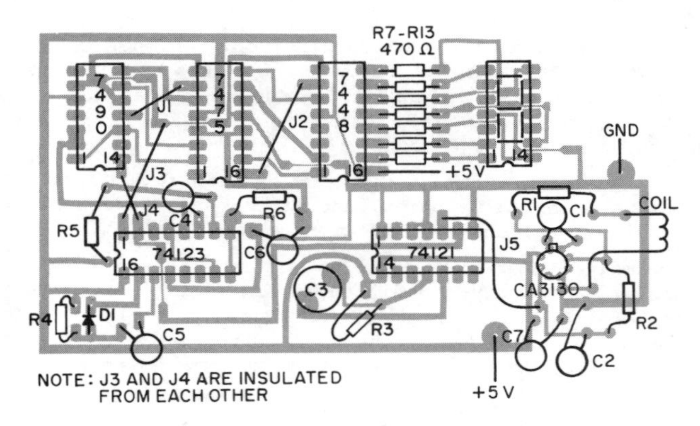

Figure 9-3 *Component Layout for Dialed-digit Monitor*

The digit displayed by the LED will remain on until the dialing of the next digit has been completed, then it will display that number and the whole process starts all over again.

Construction

While parts layout is not critical for this project, the number of integrated circuits used and the large number of interconnections, make it very necessary to use printed circuit board construction (Fig. 9-2). Also, since there are a lot of leads to be soldered on an integrated circuit, certain construction techniques should be observed. These are outlined in the beginning of Chapter 3. Read them again before continuing. There are also some special precautions that should be observed when handling IC1. IC1 is a CMOS device and has a very high input resistance and thus is very sensitive to even very small currents. Therefore do not remove IC1 from its holder until you are ready to install it. Do not store it in a nonconductive plastic tray. Make sure you use a grounded soldering iron (such as the Ungar 3-wire series) when you solder IC1. And ground any test equipment that you connect to any of the leads of IC1. If you follow these simple rules, everything will work out fine.

The integrated circuits used here all require a 5-volt power supply, so one designed according to the description of Figure 3-2 should be included as a part of this project. The whole unit, including power supply can be mounted in a small bakelite box. The display is mounted on the top via a socket that is used for integrated circuits.

Installation and Operation

To use the dialed-digit monitor, place the telephone pickup coil underneath the phone towards the right rear. Turn the monitor on and lift the receiver off the hook. Lifting the receiver should generate a pulse or two and the display may show a number. Dial zero to set the display back to zero. Now dial a number. As soon as you finish dialing, the number should appear on the LED display. If it does not, you may not have placed the induction coil in the right position. Move it around until it does appear.

10. Pulse Programmer

The pulse programmer seen in Figure 10-1 is primarily intended to be used with the cassette autodialer alarm system described in Chapter 11. It is used to program tapes which can then dial a telephone or control some other circuitry. However, a little bit of imaginative thinking will quickly make you realize that there are many other applications for which it can be used.

For example, you can combine the programmed tapes with the recorder and dialer portions of the autodialer and produce an automatic telephone dialer. A C-60 cassette used for this application could easily store one hundred phone numbers.

Another use for the pulse programmer is in synchronizing tape/ slide presentations. In this use the tones recorded on the tape are used to short out the remote control contacts of the projector and thus cause the slide mechanism to advance.

The pulse programmer can also be used to prepare tapes that will control a sequence of events, such as turning lights on and off in your house while you are away. If endless cassettes are used, things can periodically be turned on and off.

About the Circuit

The heart of the pulse programmer is a two-transistor astable multivibrator (Fig. 10-2). The frequency of oscillation of the multivibrator is determined by R1,R2,C1, and C2. By setting R1 = R2

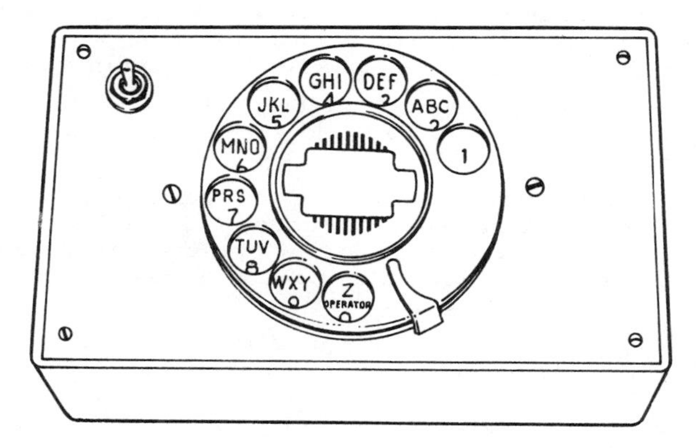

Figure 10-1 *Pulse Programmer*

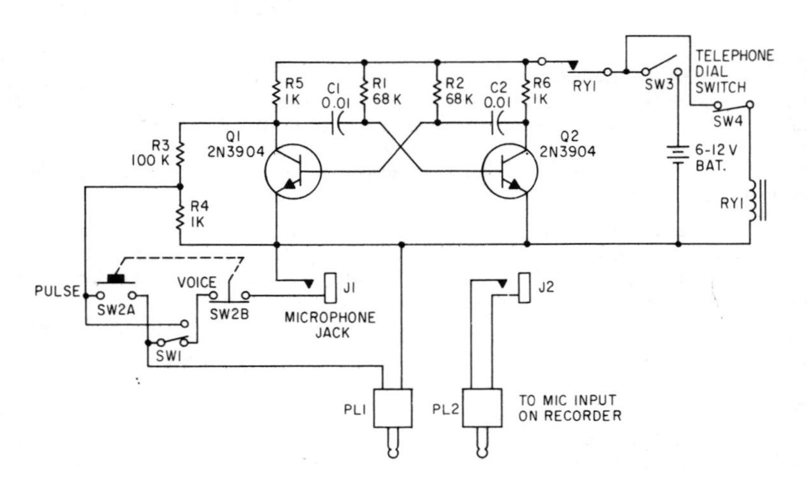

Figure 10-2 *Two-transistor Astable Multivibrator Schematic*

and C1 = C2 the frequency of oscillation is f = 1/T where T = 1.38R1C1. For the components shown, the oscillation frequency is about 1 kHz.

The astable will provide the tones that will be recorded onto the tape for controlling the relay at the output of the recorder.

Because not all tape recorders have a high impedance input, the microphone input was chosen. By connecting this to the voltage divider network R3/R4 the desired high input resistance to the astable can be achieved. The voltage divider also prevents any damage to the recorder from excessively-high oscillator voltages.

While the low-resistance input of the divider goes to the microphone input of the recorder, the high-resistance side goes to the output of the multivibrator.

To get the multivibrator to produce tone pulses that correspond to the digit pulses produced by a telephone, the astable output circuitry is controlled by relay RY1, which itself is controlled by the telephone dial. The telephone dial may be removed from any old telephone or purchased surplus.

The relay is needed because the pulsing contacts of the relay are normally closed, and normally open contacts are needed to control the output of the astable multivibrator. The relay is connected so that when the dial is operated it produces a series of tone pulses for each number dialed. If, for example, a 3 is dialed, the relay will close and open three times producing three pulses. The pulse rate of the dial, by the way, is about 10 pulses per second.

Construction

The pulse programmer is one of the simpler circuits in this book and should present no construction problems. The unit is built into a plastic utility box that measures 6¼" × 3⅝" × 2". This size box will accommodate the standard telephone dial as well as the multivibrator printed circuit board, relay, and battery.

The 3¼-inch hole in the cover plate required for the dial can easily be cut with a hole saw. Before doing this, place the plate on a wooden board and prevent it from rotating by hammering four nails through the corner holes and part way into the wood. When you start to cut the 3¼-inch hole it is best to use a variable speed drill on low, if one is available. If not, pump the trigger switch on and off so that the hole saw only makes a few revolutions at a time. By working slowly and carefully you will get a perfect hole. The dial can be mounted to the cover plate with two screws and Z brackets. (For PL2, any wire can be connected to the plug's tip or body.)

Installation and Operation

To connect the pulse programmer for use, a four-conductor shielded cable is needed. This cable should have a miniature and subminiature plug on both ends. The minature plug goes to the tape recorder's microphone input, while the subminiature one goes to the remote control jack so that the recorder can still be controlled by the remote switch generally found on the microphone.

To program a series of pulses from the telephone dial turn on SW3 and place SW1 in the pulse position and operate the telephone dial. This will feed a train of tone pulses to the recorder input. When you are finished recording pulses, flip the switch to the voice position and record your message. Additional pulses of longer duration, such as those used to separate messages in the autodialer, can be made by operating pushbutton switch SW2. This will produce an output pulse for as long as it is held down, no matter what position SW1 is in.

Parts List

R1 — 68k ohms	BAT — 6-12V battery
R2 — 68k ohms	J1 — miniature phone jack
R3 — 100k ohms	J2 — subminiature phone jack
R4 — 1k ohms	PL1 — miniature phone plug
R5 — 1k ohms	PL2 — subminiature phone plug
R6 — 1k ohms	RY1 — 12V spdt relay
C1 — 0.01 µF	SW1 — spdt switch
C2 — 0.01 µF	SW2 — dpst pushbutton nc and no
Q1 — 2N3904	SW3 — spst switch
Q2 — 2N3904	SW4 — normally closed switch from telephone dial

11. Autodialer

You're away on vacation and a burglar breaks into your home. He trips an alarm in the process but nothing seems to happen. Minutes later he is caught in the act by the police. They were called to the scene of the crime automatically by the cassette autodialer.

The autodialer is a very simple device that attaches both to your telephone and a standard cassette recorder. When a cassette tape has been prepared with the pulse programmer (described in Chapter 10) it will automatically dial any number selected and give a message of any length. If you want, several different telephone numbers, each with a different message may be dialed in sequence. This comes in very handy in places where the law prohibits burglar-alarm dialers from being connected directly to the police department. In such cases the phone numbers of several friends or relatives can be dialed in turn to make sure that at least one of them calls the police to let them know a robbery is in progress.

Dialing can be done in one of two ways, either with a pulsing relay that is in series with one of the leads of the telephone, or with a solenoid that pulses the cradle button. The first method is easiest and requires no mechanical linkage. Depending on where you live and which telephone company services you, a coupling device that interfaces the dialer to the telephone may be required as an added safety feature.

If you don't want the expense of the interface and don't want to take any chances with the phone company, then the solenoid approach is for you. With this technique the cradle button, which you normally use to hang up the phone, is pulsed by the solenoid for each number. If an 8 is dialed, the solenoid will press the cradle button down 8 times. This action accomplishes the same thing as the contacts on the telephone dial; it opens and closes the telephone circuit every time a pulse is applied.

About the Circuit

The automatic cassette dialer (Fig. 11-1) can be broken down into three main parts: (1) the recorder; (2) the control circuitry; and (3) the dialing circuitry.

The recorder can be any cassette recorder, for all of the information that is recorded and played back on the machine is well within the frequency response of even the least expensive device.

The control circuitry interfaces the unit to any already existing burglar-alarm system. Almost all burglar-alarm systems are normally closed alarm switches for increased reliability. When any switches are opened, the gate of the SCR becomes disconnected from ground and

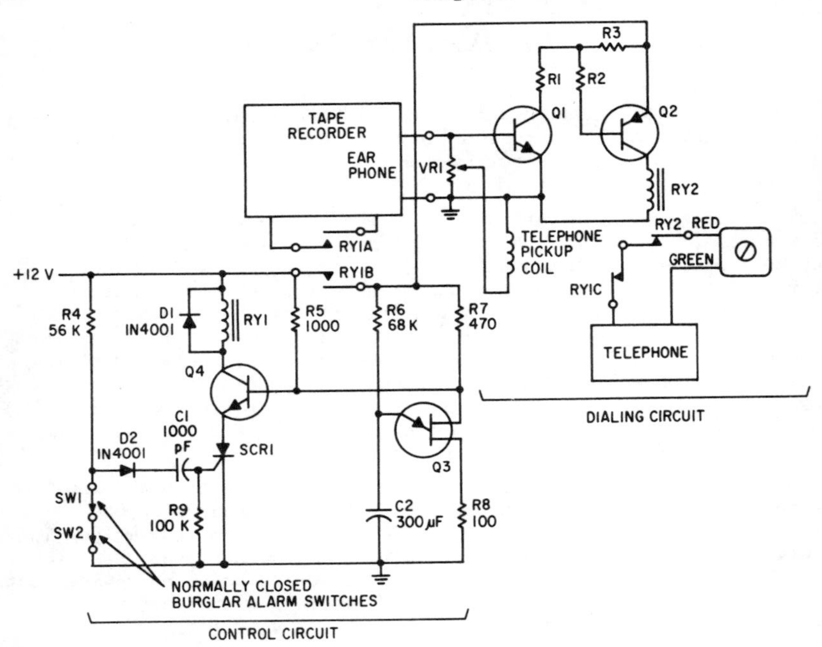

Figure 11-1 *Automatic Cassette Dialer Schematic*

a positive trigger signal is applied to the gate of the SCR, turning it on. The normally closed alarm switch only has to open up momentarily for the dialer to start. After the SCR is triggered the gate has no control over the device and even if it is returned to ground potential, it will not turn off the dialer.

Parts List

C1 — 1000 pF	Q3 — 2N2646
C2 — 300 µF 16V electrolytic	Q4 — 2N3904
R1 — 12k ohms	BAT — 12V battery
R2 — 1k ohms	SW1 — spst switch (magnetic reed)
R3 — 2.2k ohms	SW2 — spst switch (magnetic reed)
R4 — 56k ohms	RY1 — 3pst 12V relay
R5 — 1k ohms	RY2 — spdt 12V relay coil resistance 250 ohms
R6 — 68k ohms	D1 — IN4001
R7 — 470 ohms	D2 — IN914
R8 — 100 ohms	SCR1 — GE 106Y
R9 — 100k ohms	VR1 — 20 Ω wire wound Lafayette 99F61384
Q1 — 2N3904	
Q2 — 2N3906	

Once the SCR is triggered, it causes relay RY1 to close. This applies power to the dialer recorder, and the disconnect timer. It also closes a set of normally open contacts that are in series with the red wire from the telephone.

The recorder starts and plays back the preprogrammed tape. The output of the recorder is fed to a ring-type induction coil and a pulse detection circuit that opens and closes a relay (RY2) for each pulse. The coil is used to induce audio signals from the recorder into the telephone line.

RY2 has a set of normally closed contacts that is in series with the normally open contacts of RY1 and the telephone. When RY2 opens and closes according to the pre-recorded pulses, it is dialing the telephone.

If the solenoid dialing method is used, the RY1 and RY2 series circuit is placed in the power lead to the solenoid, so that it will pull-in and let-out with each pulse.

After the series of dialing pulses is completed, the audio message appears at the output of the recorder. The message will not key relay RY2 because its volume, and hence the voltage generated, is lower than that of the dial pulse. The volume control on the tape recorder is adjusted so that the relay closes when it should. The level of the sound induced into the phone is controlled by VR1. The message is induced into the telephone via the coil and heard at the other end.

If more than one phone call and message is to be made, each message except the last one will be followed by a tone that is between 5 and 10 seconds long. This tone will cause relay RY2 to open the telephone line long enough for it to disconnect and prepare the telephone for the next call.

After the disconnect tone the dialing pulses for the next call repeat the same way as in the first. When the time delay determined by $T = R6C2$ has elapsed, the timer produces a pulse which turns off transistor Q4. The high impedance of the emitter-collector circuit in the nonconducting transistor greatly reduces the current through the coil of relay RY1, causing it to unlatch. This disconnects power from the autodialer and opens the phone line circuit.

The reset timer is simply a unijunction transistor oscillator with a low frequency. The trigger pulse is taken from B2. This provides the required negative-going pulse to turn off Q1.

Construction

The construction of the autodialer will be largely determined by the method of dialing used. If the series relay is used, the whole unit may be enclosed in a small bakelite case. If the solenoid dialing scheme is chosen then a 5" × 9" × 2" chassis like the one used for the remote ring indicator is suggested. This will allow the phone to sit on top of it and provide a firm foundation for the vertical arm that will hold the pulsing solenoid.

The vertical arm should be 10 inches high and is easily constructed from a piece of wood 2 inches wide and 1 inch thick. To position the supporting arm, place the telephone on the chassis the way it normally will be placed. Then mark the spot where the arm is parallel with the receiver cradle. Fasten the vertical support with two screws.

Next, place the solenoid against the support and let the plunger hang down. Lower the solenoid until the plunger depresses the cradle button, mark the spot on the support, and fasten.

With the mechanical work out of the way, the rest of the device is relatively simple to fabricate. The circuit can be built on perf board or on a pc board made from the pattern supplied.

The unit has been designed so that it can work from a 6V lantern battery that is generally used in alarm systems, but can also be used with the power supply described in Chapter 3.

Installation and Operation

If the solenoid dialing method is used, installation simply means placing the telephone on the chassis, taking the receiver off the hook, and allowing the solenoid plunger to hold the cradle button down.

If the series-relay dialing method is used, one of the leads from the telephone, preferably the red one, is connected in series with the contacts of RY1 and RY2. In both cases, the telephone pickup coil is then placed on the earpiece of the handset of the telephone. Before placing the unit into operation, play the entire message tape and time it. Add 5 seconds on to this time and then set the unijunction oscillator so that it will produce a pulse after that amount of time has elapsed. This is done by adjusting the value of R1 until the desired time delay is obtained.

You are now ready to test the cassette autodialer. Connect the unit to your burglar alarm system so that in the untriggered position A and B are normally shorted. Now momentarily depress test switch SW1. This simulates a burglar entering the protected area and triggers the autodialer.

If the dial pulses that are recorded on the tape do not cause the relay to operate, the volume is too low. It may be necessary to rerecord them at a higher volume.

If the voice message causes the relay to kick in and out, the volume is too high. Lower the volume or rerecord the message.

12. Dialer

While the telephone dialer described in the previous chapter can be used for the everyday dialing of telephone numbers, it is best for burglar alarm applications. To use it as an everyday dialer you need a recorder that has a footage indicator so that you can easily locate the number you need.

If your main wish is to relax and not strain your fingers while dialing, then this mechanical telephone dialing unit is for you. Like the automatic dialers that are available from the telephone company, this one uses an individual card for each phone number. This makes storage and location of numbers easy and convenient.

Unlike telephone company units, however, this one uses a disc shaped cardboard card for each number. Programming of telephone numbers is as simple as punching a few holes; and operation of the dialer is as simple as putting a record on your phonograph.

Like the dialer described earlier, this one can be operated indirectly using a solenoid and the telephone cradle switch, or by making a direct connection to the phone line. But remember, if a direct connection is made to the phone line, a telephone company provided interface may be needed.

About the Circuit

The circuit of the dialer is very simple as you can see from Figure 12-1. The entire device is constructed from a maximum of six components. They are pushbutton SW1, microswitches SW2 and SW3, toggle switch SW4, motor M1, and if the indirect dialing approach is used, solenoid S1.

In operation, a cardboard disc with the desired phone number is placed on the turntable so that the center and key holes of the disc fit over the pins on the turntable.

To dial a number, SW1 is pressed momentarily. This causes the motor to operate and the turntable begins to rotate. As soon as it has moved away from microswitch SW2, SW2 closes its normally closed contacts and supplies power to the motor, allowing it to complete exactly one rotation. At the end of its rotation the motor is shut off by SW2, which is caused to operate by the bottom part of the keypin.

The actual number dialing is performed by the second microswitch SW3 when using a direct connection to the phone line, and by S1 when not.

For direct line operation, SW3 is connected via its normally open contacts to the red wires of the telephone and the telephone line.

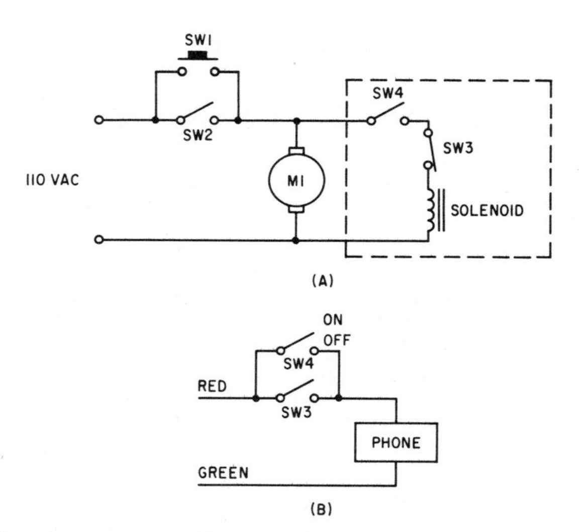

Figure 12-1 *Disc Dialer Schematic Using Solenoid to Dial* (A). *Boxed-in Area is Eliminated and Connection is Made to Phone Line in Direct Approach* (B).

This puts SW3 in series with one leg of the phone circuit. Once the dialing disc is properly inserted, SW3 has its normally open contacts held closed by the edge of the disc.

As the disc rotates, holes punched on the edge of the disc permit SW3 to open and close, pulsing the telephone line and dialing the number. SW3 opens and closes once for each hole punched on the disc edge.

For indirect dialing, the normally open contacts of SW3 are used to operate a solenoid. As the disc rotates, the contacts open and close, causing the solenoid to drop down and pull in.

Parts List

M1 — 6rpm motor, 110vac
SW1 — spst no pushbutton
SW2 — spst microswitch
SW3 — spst microswitch
SW4 — spst toggle switch
SOL — 110vac solenoid

For direct dialing, SW4 is needed to short SW3 when the dialer is not in use. When the indirect dialing method is used, SW4 is closed before pushbutton SW1 is pressed. This pulls in the solenoid and prepares the phone for indirect dialing.

Construction

Most of the construction of this device involves mechanical work. The first thing to do is to make the turntable that will hold the cardboard dialing discs. This is done by cutting a 4½-inch circle out of a piece of ¼-inch plywood. The turntable can be cut with either a hole saw or with a sabre saw.

After the turntable is cut out, drill a ½-inch hole in the center of the disc and fasten a piece of ½-inch dowel to the turntable with glue. This will be the drive shaft. The length of dowel will depend upon the depth of the box you are mounting the dialer in.

The bottom end of the dowel (the one not connected to the turntable) should have a ¼-inch deep slot sawed into it right across the diameter of the dowel. This will allow the dowel to lock onto the motor shaft and rotate the turntable. Next cut a piece of 2" × 1" wood 10 inches long. This is the vertical supporting arm for the solenoid and is not needed if a direct connection to the phone line is used. Mount it at a point on the chassis where it will be parallel to the telephone cradle. Fasten it to the chassis with two screws and mount the solenoid on it so when the plunger is extended, it holds the cradle switch down.

When this is done, locate a point 1½ inches from the center of the turntable and drill a small pilot hole. Mount a screw there so that the head of the screw is on the underside of the turntable. This will act as a key for the dialing discs.

Now drill a ⅝-inch hole in the center of the cover of the box the unit is to be mounted in. This is for the drive shaft. Mount the motor directly underneath this hole and lock the drive shaft onto it by simply pushing it down. Next mount microswitch SW2 so that its roller is held down by the turntable key. SW2 should be positioned so that when the turntable rotates and the key moves away, the microswitch will close.

Microswitch SW3 is mounted so its normally open contacts are held closed by the edge of the disc, and open when the roller comes to a punched hole on the edge of the dialing disc.

The easiest way to prepare dialing discs is to make photo copies of the full size disc shown in Figure 12-2 and paste these copies onto cardboard. Then cut out the discs and make holes for the center pin and keypin. To program the card, simply take a hole puncher and punch out the appropriate number in each of the seven number fields. For example, if the first digit of the number you want to dial is 3, you punch out the first three holes in that number field.

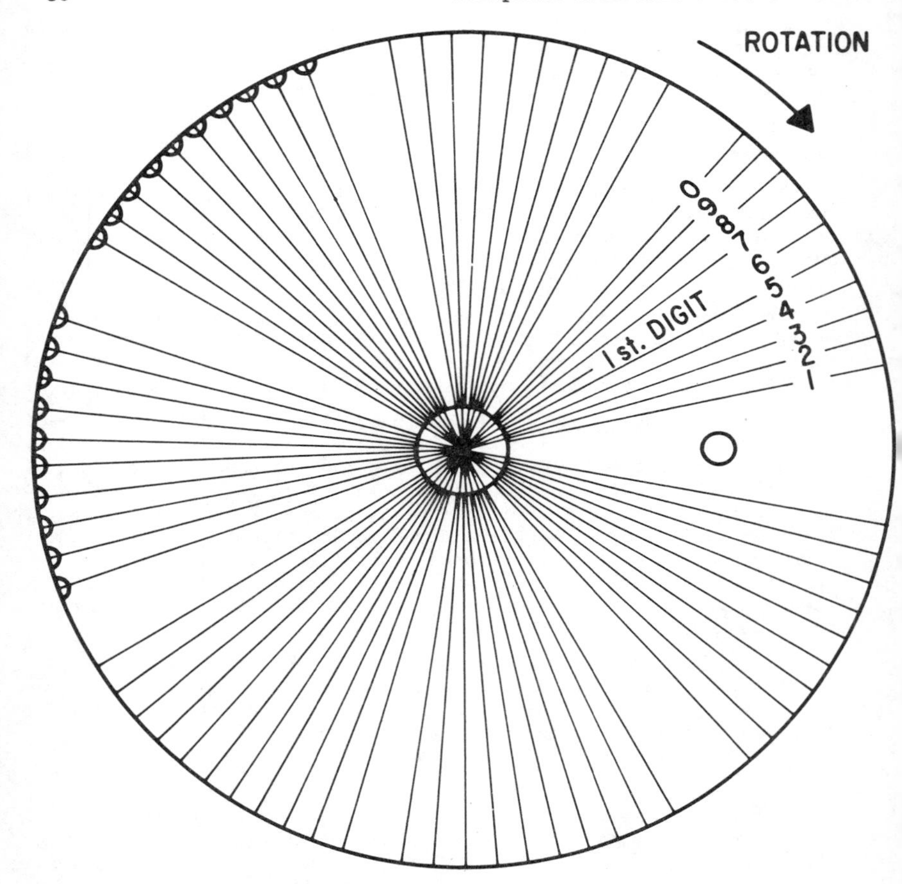

Figure 12-2 *Dialing Disc (Actual Size)*

Installation and Operation

As with the autodialer, installation will vary with the type of dialing scheme chosen. If the direct approach is used, the wires coming from the parallel combination of SW3 and SW4 are connected in series with the red lead on the telephone and the red wire coming from the telephone junction box.

For cradle-switch dialing, mount the solenoid so that when the plunger is extended, it holds the cradle-switch button in the on-hook position. Next, insert the disc that has been programmed to dial the number you are interested in. Be sure that you engage SW3 in the process.

With the unit connected to the ac voltage, you are now ready to try it out. Turn on switch SW4. Press pushbuttom SW1. The motor should start to rotate. When you let go of the switch, the motor should continue rotating until it has completed one revolution. If it does not stop, check SW2; it is probably not located in the right position for the keypin to activate it.

While the disc rotates, SW3 should be pulsed on and off. If it doesn't pulse then it is either not located close enough to the dialing disc, located too close to the dialing disc, or the dialing disc is not perfectly circular.

Once you get used to using the dialer, you'll wonder why you ever did without it. One final suggestion, the dialing discs can be stored in file cabinets that are available from most stationary stores. These cabinets usually come with alphabetical index separators.

13. Telephone Burglar Alarm

In earlier projects you learned how to build an automatic telephone burglar alarm and tape programmer which could be used to dial a specific number in case of a burglary. If you want a less sophisticated device that can do the same job — that is, monitor a remote location for a possible break-in — then you can use this remote telephone burglar alarm.

With this device hooked up to the telephone at the place being guarded and the telephone at your monitoring location, you know immediately if anyone has broken in to the area being guarded. The unit has two basic parts: (1) the remote alarm which is connected to the premises being protected, and (2) the central monitor which connects to your phone and listens for a burglar-alarm signal from the remote location.

About the Circuit

The remote unit (Fig. 13-1) that is located on the premises to be monitored, consists of an alarm circuit and a tone generator which is activated when the alarm is tripped.

The alarm circuit accepts both normally-open and normally-closed switches. It also has provision for triggering by a touch switch.

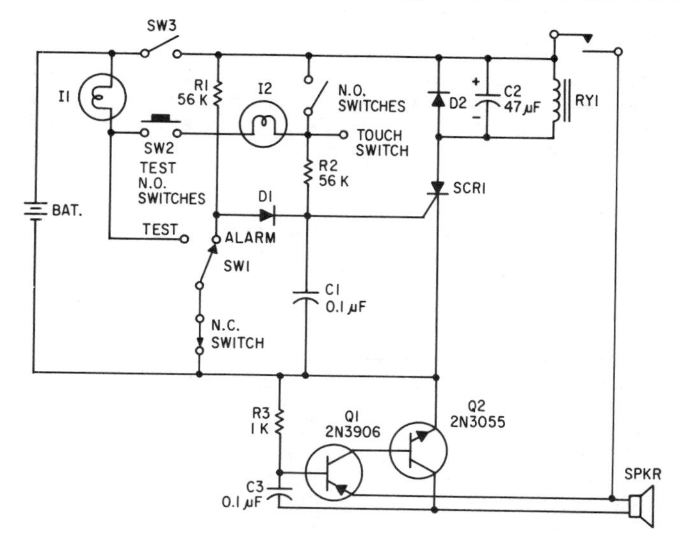

Figure 13-1 *Remote Alarm Schematic*

For this mode of operation, the metal object to be protected, such as a door knob, is connected to the circuit by a single wire.

In operation, when a normally closed sensor is opened by an intruder, current flows through diode D1 triggering the SCR and triggering the alarm. Alternately, if one of the normally-open switches closes, current flows through that switch to the gate of the SCR to trigger it and again the alarm is set off.

The alarm system also contains circuitry to test the alarm switches used. For normally-closed switches it is only necessary to switch SW1 to the TEST position. Lamp I1 should come on. For normally-open switches it is only possible to tell if they are shorted. This is done by switching SW1 to TEST and pushing button SW2 momentarily. I2 should not light.

Instead of triggering a bell or a siren when the alarm is activated, this circuit triggers a tone generator which can be acoustically or inductively coupled into the telephone.

The tone generator consists of a keyed complementary astable multivibrator. This tone generator is one of the simplest ones around and consists of two transistors, a resistor, a capacitor, a speaker, and a power source. Parts values aren't critical, just about any value capacitor and resistor will work. Q2 however should be a power transistor so that it can handle the current required to drive the speaker.

When the burglar alarm goes off, the SCR is triggered and causes a voltage to appear at the base of transistor Q3. This voltage turns Q3 on and turns the astable oscillator on. C1 then charges up through R1. The voltage build up on C1 turns Q1 on, supplying current to the base of Q2 so that both transistors conduct. When Q2 turns on however, C1 shorts out and both transistors turn off. C1 then recharges and the whole process repeats itself at a rate determined by R1 and C1.

Every time Q2 conducts current-flows to the speaker, an output tone is produced. The speaker is placed next to the mouthpiece of the remote telephone so that the alarm signal is sent via telephone to the monitoring location where it can be detected and set off a warning device.

Parts List

R1 — 56k ohms	Q1 — 2N3906
R2 — 56k ohms	Q2 — 2N3055
R3 — 1k ohms	RY1 — 6V relay spst
C1 — 0.1 μF	SCR1 — GE 106Y
C2 — 47 μF 15wvdc electrolytic	SPKR — 8 ohm speaker
C3 — 0.1 μF	SW1 — spdt switch
D1 — 1N914	SW2 — spst pushbutton (momentary)
D2 — 1N914	SW3 — spst switch
I1 — 6V lamp	MISC — normally open and closed alarm switches and 6V battery
I2 — 6V lamp	

At the monitoring location a sound switch (Fig. 13-2) is used to detect the alarm signal and set off the central alarm. This sound switch is the same type as described in Chapter 4 to detect the ring signal of the telephone.

A crystal microphone is placed next to the earphone of the monitoring station. When the alarm signal is sent, the microphone picks up the sound and feeds it to the base of transistor Q4. Q4, which is normally conducting, and thus shorting out capacitor C2, turns off, allowing charge to build up on the capacitor. This creates a positive voltage which then triggers SCR2. The SCR can be used to trigger a bell or buzzer as indicated, or the circuit in the dashed box can be connected at the X marks so that the SCR triggers an electronic siren.

Like the alarm tone generator in the remote location, the siren is a type of complementary multivibrator. This one, however, produces a much louder piercing sound. It will yield best results if it is used with a horn-type speaker.

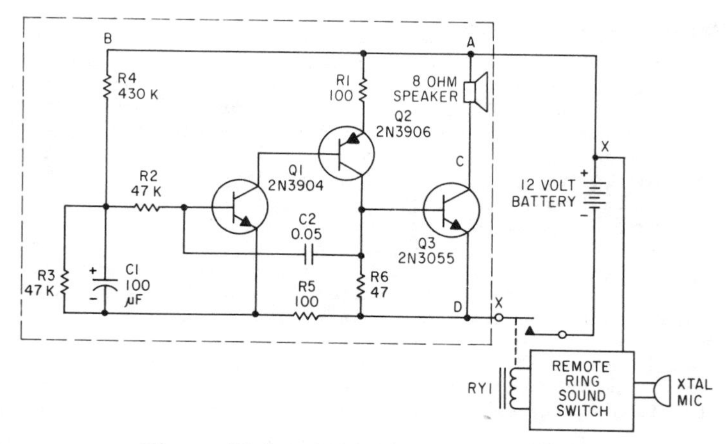

Figure 13-2 *Central Unit Schematic*

Parts List

R1 — 100 ohms	C2 — 0.05 µF
R2 — 47k ohms	Q1 — 2N3904
R3 — 47k ohms	Q2 — 2N3906
R4 — 470k ohms	Q3 — 2N3055
R5 — 100 ohms	SPKR — 8 ohm speaker
R6 — 47 ohms	SW1 — spst switch
C1 — 100 µF 16wVdc electrolytic	BAT — 12V battery

Construction

The remote unit with the alarm and tone-generator circuitry is built into a box that measures 10" × 4" × 2½". A wooden cheesebox will do nicely here. This size box will allow you to place the telephone receiver into it and make it handy to couple the speaker to the phone. If such a box is not available it will be necessary to find some other way (such as using a few heavy rubber bands) to hold the speaker next to the handset's mouthpiece. The power source should be an external lantern battery or an ac powered dc supply (see text description of Fig. 3-2). Bring points A,B,C, and D to terminal posts on the box. This will facilitate connection to the burglar alarm switches.

The central monitoring unit can also be built into the same type of box as the remote unit. Everything except the alarm bell or speaker can be housed in this box. The only external connections needed are for power and the alarm device.

Installation and Operation

To install the remote unit wire up the normally-closed and normally-opened alarm switches and any metal objects that are to be protected by the touch switch. Make the appropriate connections to the box. Touch switch connections are made to terminal D. After the unit is wired up, and with the master switch still off, flip switch SW1 to the test position. I1 should light up. If it doesn't, one of the normally-closed switches is open or not making good contact. Next, press switch SW2. If I2 lights up, one of the normally-open switches is either not functioning properly or is shorted. Once the switches have been checked, make sure the tone generator works. Turn the master switch on and touch terminal D. This should trigger the alarm and turn the tone generator on. After you've made sure that the remote unit works you can set it up for use. Dial the phone number of the monitoring location. When someone answers tell him that you are ready to arm your alarm system and that he should place the telephone receiver next to the crystal microphone.

To test the central monitor, set off your alarm so that the tone generator is sending out its signal. The sound switch at the other end should trigger as soon as it hears the tone. If it doesn't, the threshold for triggering the SCR is too high and should be lowered by adjusting potentiometer R2. Be careful not to make the threshold too low because often there are spurious noises on the phone lines which can trigger the alarm.

With the threshold properly set, the tone from the remote location should set off the alarm at the central location.

While this system obviously ties up your telephone while it is in use, it saves the cost of renting a private line from the telephone company.

14. Answering Machine

Have one message you want to give to all the people who call you? Do you need a telephone answering machine too? Then build this dual-purpose automatic message repeater and answering machine. With the switch in the repeat mode you will be able to have the unit answer the telephone, give a repetitive message to all those who call — such as this week's meeting has been postponed until Thursday — and then hang up. In the answer mode the machine will give your message of any length to the calling party and then take his message of any length for you, a feature that even some of the most expensive machines do not have.

Since all of this is done inductively you do not have to worry about the telephone company objecting to your machine.

Externally the unit looks like the cassette autodialer described in Chapter 11. That is because it is built in the same type of housing as that unit, and activates the telephone in the same way, by using a solenoid to lift up the cradle button (Fig. 14-1).

About the Circuit

Depending on whether you want a message repeater only or a machine that can both take messages and give them, you will need one or two cassette recorders.

If you need only a message repeater, the circuitry for the unit can be somewhat simplified, and you'll need only one recorder.

There are two basic circuits used for control in the machine. The first you have already become familiar with in the remote ring indicator and the cassette autodialer. This is the sound switch (or ring detector) and it is used to detect the ringing of the telephone. The circuit is identical to that used in the remote ring indicator except that the additional reset circuitry (Q1 in Fig. 4-1) used in the indicator is not used (Fig. 14-2).

As you can see from the figure, there has been one slight addition to the original ring detector that will make it possible to allow the machine to hang up after it has given a message and before it has taken one, making it a message repeater only.

The second basic circuit is a silence detector (Fig. 14-3). The silence detector is the circuit that makes this machine better than expensive ones. It detects when there is a period of silence that exceeds the desired amount of time and then produces a pulse which switches the machine.

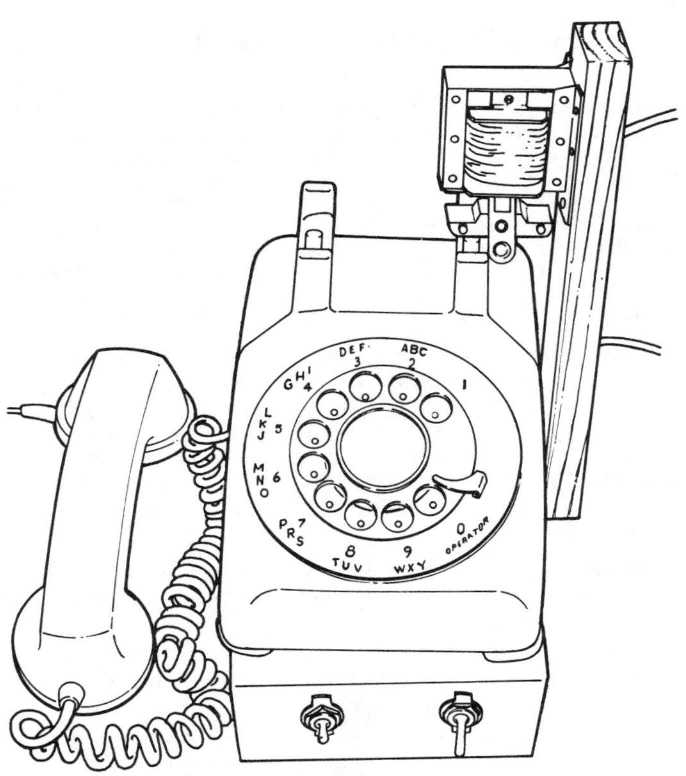

Figure 14-1 *Telephone Answering Machine*

For the message repeater you will need one silence detector, while for the complete answering machine you'll need two of them.

Operation of the silence detector is simple. Output transformer T1 is connected backwards so that it will raise the output voltage of the tape recorder. It connects to the earphone jack of the recorder. After it goes through the transformer, the signal is then filtered by R1 and C1 to eliminate any high-frequency components from the erase head oscillator. The signal is then rectified by diode bridge DB1.

The pulsating dc voltage derived from the bridge is then applied to the emitter of unijunction transistor Q2 through capacitor C2. When there is an audio signal present at the input of transformer T1, the negative voltage generated builds up on C3 and prevents the

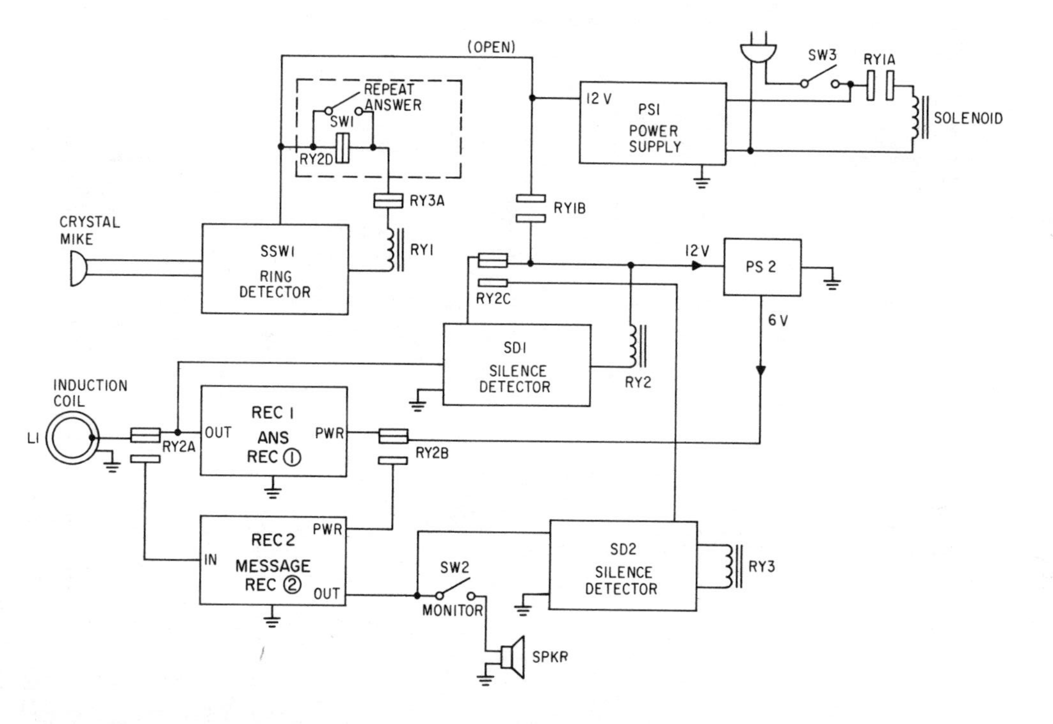

Figure 14-2 *Telephone Answering Machine Schematic*

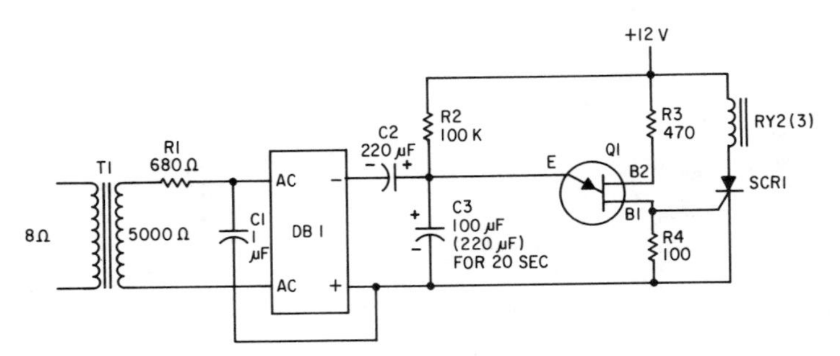

Figure 14-3 *Silence Detector Schematic*

unijunction oscillator from firing. When the audio signal ceases, charge builds up on C3 until it reaches the firing voltage of the unijunction. At the firing voltage the unijunction turns on and discharges C3, whereupon the cycle repeats itself. The time between the end of audio and the output pulse is roughly determined by $T = R2 \times C3$. The values shown for the two silence detectors yield a delay of 10 and 20 seconds respectively.

The first silence detector is attached to the output of recorder 1, the answer recorder. The second silence detector is connected to the output of recorder 2, which records the caller's message.

Parts List

Answering Machine	Silence Detector
L1 — ring type telephone pickup coil	R1 — 680 ohms
PS1 — 12V power supply (see Fig. 3-4)	R2 — 100k ohms
PS2 — 6V power supply (see Fig. 3-4)	R3 — 470 ohms
RY1 — dpst 12V relay	R4 — 100 ohms
RY2 — 3pdt 12V relay	C1 — 1 µF non polar
RY3 — spdt 12V relay	C2 — 220 µF 16Vdc electrolytic
SD1 — 10 sec silence detector (see Fig. 14-5)	C3 — 100 µF 16Vdc electrolytic
SD2 — 20 sec silence detector (see Fig. 14-5)	C4 — 220 µF 16V electrolytic
SPKR — 8 ohm speaker	DB1 — diode bridge, 50v piv
SW1 — spst switch	Q1 — 2N2646 unijunction
SW2 — spst switch	RY2 — 3pdt 12V relay
SW3 — spst switch	RY3 — spdt 12V relay
XTAL — crystal microphone	SCR1 — GE 106Y SCR
MISC — two cassette recorders	T1 — 8 ohm-5000 ohm plate transformer

The machine works in the following sequence. When the machine is on and you get a telephone call, the sound switch detects the ring and closes relay RY1. This does two things. First it applies power to the solenoid so that it pulls in and allows the cradle switch to close, answering the phone. At the same time another set of contacts on RY1 apply power to recorder 1 and to the first silence detector circuit.

Recorder 1 plays back your message to the caller on an endless cassette. Since the silence detector for this recorder is set to go off about 10 seconds after the message stops, there should be about 15 seconds between the end and the beginning of the message. This will provide enough time for the detector to trigger when the message is finished but not give it enough time to trigger when it is activated for the next call.

At the end of the message Q1 fires, producing a positive going pulse at B1 and a negative going one at B2. If switch SW1 is open, RY2 is momentarily opened, which opens RY1 and the unit is ready for the next phone call. This is the message repeater mode of operation where the machine gives the message and hangs up.

If SW1 is closed RY2 latches on. This switches the 6 volt power supply from recorder 1 to recorder 2. Another set of contacts on RY2 switches power from silence detector 1 to silence detector 2, while a third set of contacts switches the induction coil from the output of recorder 1 to the microphone input of recorder 2.

Recorder 2 starts and records the caller's message. About 15 seconds after the caller finishes his message the second silence detector fires turning RY1 off, and resetting the whole machine for the next phone call.

Construction

There are a few things that you must pay attention to when you are constructing this project. The first is that telephone pickup coil L1 must be a circular type device that fits around the handset's receiver. A flat coil that goes underneath the phone or one of the suction cup devices will not work because while they will pick up audio from the phone, they cannot be used to induce signals into the phone. Because of its construction the circular pickup coil specified can do both.

If it is impossible to get this coil in your area you may eliminate the RY2 contacts associated with that coil and permanently wire a speaker to the output of recorder 1 and any type of pickup coil to the input of recorder 2. The speaker, however, must be placed near the mouthpiece of the telephone so that its output will be acoustically coupled to the telephone.

Connections between the recorders, inductive pickup, and the control circuitry can be made by wires with plugs on both ends. If a metal chassis is used however, be sure to isolate the jacks from the chassis by using insulating hardware or by mounting all of the jacks on a piece of plastic and the plastic to the chassis. If you do not do this, undesirable ground loops will not allow the unit to function properly. All interconnections that carry audio signals should be made via shielded cables to avoid 60 Hz pickup. This includes connections to the silence detectors and the monitor speaker.

The power supply for the answering machine consists of two regulated supplies. One is a 12-volt supply for the control circuitry and the relays. The second is a 6-volt supply for the recorders. This eliminates the need to worry about batteries running out.

If your recorders require a different voltage it is only necessary for you to change the value of the zener diode D in Figure 3-4 to the appropriate value.

Since the 6-volt supply provides power to only one recorder at a time, it does not dissipate a lot of heat and the power transistor does not need a heat sink.

The 12-volt supply, however, dissipates a lot of heat and therefore requires a heat sink for the power transistor. The simplest way to heat sink the device is to mount it on the aluminum chassis. When doing this however it is necessary to make sure that you insulate the transistor from the chassis so that it does not make electrical contact with it. Use a mica washer (available at most parts suppliers) to insulate the transistor from the chassis. Also make sure all drilled holes are smooth and no burrs are present.

The last important construction point for this device is: Be sure you mount the chassis on rubber feet. This is needed because when the solenoid drops to hang up the telephone it causes the chassis to move a little. This slight movement causes a noise that can trigger the sound switch. If you use rubber feet however the unit will not move and therefore will not cause any false triggering.

Installation and Operation

To test the unit after it has been constructed, place the telephone on top of the chassis, remove the handset, and allow the plunger of the solenoid to drop down and hold the cradle switch in the on-hook position.

Now using one of the recorders, take a 30-second endless cassette (longer if your message is longer) and record the message you want a person calling you to hear. A typical message would be:

> You have reached the home of Joe and Mary Smith. We cannot come to the phone now but would like to call you back as soon as possible. Will you therefore leave us your name and telephone number. If you have a short message you'd like to leave, please do so. This

> recorder will shut off automatically after you've finished talking. There will be no beep, so please leave your message now.

This message will take about 15 seconds. If you're using a 30-second cassette it will leave just the right amount of time between the end and the beginning of the message.

After you've made the message, connect both tape recorders to the machine. Put recorder number 1 in the playback mode and recorder number 2 in the record mode. Now have a friend call you. After the first or second ring the sound switch should trigger and close relay RY1. This will cause the solenoid to lift and start recorder 1. You should be able to hear your message being played in the telephone earpiece.

If SW1 is in the repeater mode the unit should shut off about 10 seconds after your message has been given. If it is in the answer mode, after 10 seconds recorder 2 should start. At this point you can listen to the person calling and decide whether or not you want to talk to him by simply flipping SW2 to the monitor mode. This connects a speaker to the output of recorder 2. If you want to talk you simply lift up the phone and shut off the machine. If you do not want to, monitor your calls then flip SW2 to mute and disconnect the speaker.

After the caller leaves his message the machine will turn off and be ready for the next call.

15. Silencer

How many times have you received a phone call at 2 or 3 A.M. that turned out to be a wrong number? Really annoying isn't it? You can prevent that of course by taking your receiver off the hook. If you have several phones in the house and want to prevent the phone in the bedroom from waking up your wife this technique can be very effective. But what if you're expecting an important call? Well now you have a choice: Let your wife be awakened by the call you're expecting or prevent that call from being completed by taking your phone off the hook. If you build the phone silencer, you'll have a third and more desirable choice, just turn off the bell on the phone in the bedroom. Commercial units that do just that are sold in many electronic equipment stores, but the price is pretty steep for a connector, switch, and adequate wire. You can build your own for a fraction of the price.

Like the phone lock, this device requires direct electrical connection to the phone line. But also like the phone lock, the silencer will cause no damage to the phone. However, as with all devices that are connected directly to the phone line, you should first check with your local telephone company to see if any interface device is needed.

About the Circuit

The circuit for the phone silencer is probably the simplest circuit in the book. It merely consists of two wires and a switch. To understand how it operates, let's look at how a telephone rings.

To ring a telephone, an ac signal of about 90 volts at roughly 20 Hz is sent along the telephone wires connected to your phone. If you take a look at the wires coming from your phone you will see that there are generally three wires, a red, a green, and a yellow. Sometimes there are additional wires coming from the phone. These are used to supply voltage for lights or other things and should not concern us here. We are only interested in the red, green, and yellow wires.

Most private phones, excluding party lines, have the yellow and green wires connected together. This is done so that the ring signal that appears on the green wire can be transferred to the yellow wire and used to ring the telephone bell. If the yellow wire is disconnected, the phone will function properly in all ways but one. It will not ring.

We use this fact to build the silencer. When the telephone company checks to see if you have extra telephones on your line, and they often do, they apply a voltage to your line to see how much current is drawn. If your line draws more current than it should according to their records, the phone company may assume you have extra phones on the line and ask you to remove them. The current they measure is the current associated with the bell and ringer circuitry. If the bell is disconnected, there is no additional current indicated.

Construction

To make your silencer, take a piece of lamp cord and an inline switch that can be mounted on it. This is the type of switch that is found on TV remote control cords. Connect the switch so that when it is on, it shorts together the two wires going to it. That's all there is to it.

Installation and Operation

To install your phone silencer, open up your telephone junction box (Fig. 15-1). If the green and yellow wires are connected to the same terminal disconnect the yellow one and put it on one of the two empty spare terminals. Connect one wire from the silencer to the green terminal and the other to the yellow terminal. Replace the junction box cover and locate the switch of the silencer at a convenient location.

You are now ready to try out the silencer. Have someone phone you. Make sure that the silencer is in the off position. As soon as you hear the phone start to ring, switch the silencer on and see how quickly the phone becomes quiet. If it doesn't, you probably have made a mistake in the way you hooked it up. Check it over again.

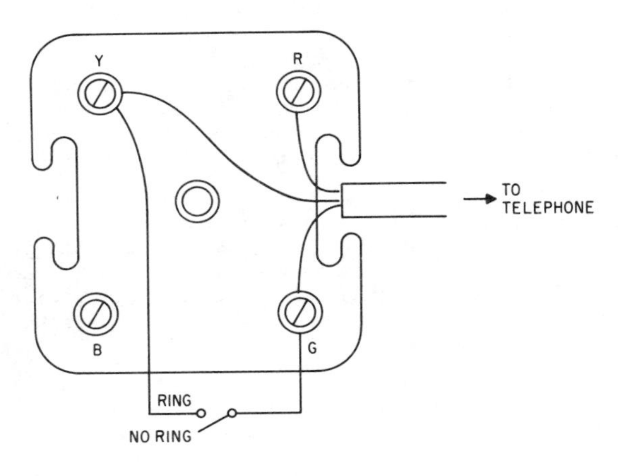

Figure 15-1 *Telephone Silencer Schematic*

16. Phone Lock

To prevent unauthorized people from using the telephone, people often go out and buy a telephone lock to keep the dial from operating. If you're one of the many people who has done that you probably know that just when you want to put the lock on, you can't find it. It's rather small and easy to lose. So is the key. Ever put the lock on and then lose the key? Try to use the phone then!

You can eliminate these problems by building and attaching an electrical phone lock to your telephone (see Fig. 16-1). While preventing unauthorized persons from making phone calls, the phone lock will not in any way affect incoming calls. The best part about the lock is that there is no key to lose. To use the phone, dial the phone lock in the proper combination and the phone is unlocked. If you want to make the combination more difficult, all you have to do is add additional switches.

The phone lock is one of the few projects in this book that requires a direct electrical connection to the telephone. The connections you make however will in no way damage the phone and cannot be detected by the phone company. In certain areas of the country, the attachment of foreign objects to the phone line is prohibited by tariffs of the local operating companies. However, recent pressure from telephone accessory manufacturers has caused the phone companies to ease up on these tariffs in many instances and to allow some devices to be connected to the line. To be sure, you should contact your local phone company.

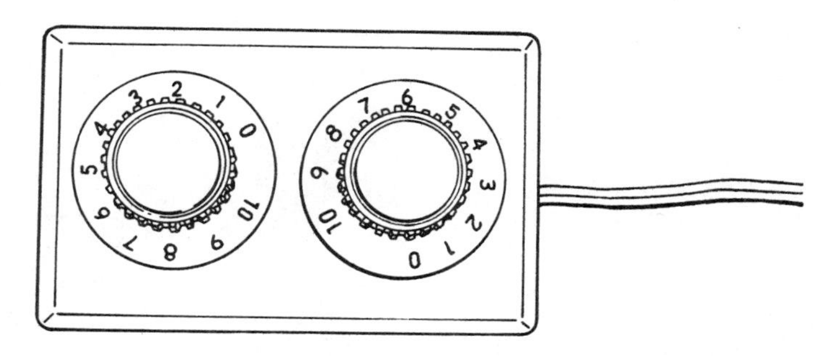

Figure 16-1 *Electrical Phone Lock*

About the Circuit

The circuit for the electrical phone lock is a very simple parallel circuit of three rotary switches (Fig. 16-2). The connection of the switches is designed so that unless the proper combination is dialed in, a short circuit will appear across wires A and B. But, when the correct combination is set on the three rotary switches, an open circuit appears across A and B.

To understand how this helps lock the phone, let's take a look at the dialing action of a telephone. Look at the back of a telephone dial. You will see that there is a set of normally-closed switch contacts that open and close as a number is dialed. The number of times they open and close depends on the number dialed. For example, the number 3 would cause the contacts to open and close three times.

This opening and closing of the contacts is what enables you to dial different numbers. When the phone is not in use and waiting to receive a call, these contacts are normally closed. So if you contact a shorting wire across the terminals of these contacts you have caused no change in the circuit.

If you try to dial the telephone while the terminals are shorted nothing will happen. Although the moving dial is causing the switch to open and close, the wire you connected across the switch makes it look as if that switch is still sitting there doing nothing. But if you were to disconnect that shorting wire, telephone dialing would proceed as normal.

This in essence is what the phone lock does. Wires A and B are connected to the normally-closed dialing switch. When the correct combination is set on the rotary switches, the circuit between A and B is an open circuit and calls can be made. But, if the rotary switches are set to anything but the proper combination, a short circuit will appear across A and B. This short circuit will prevent the dialing switch in the telephone from pulsing the phone line and dialing a number.

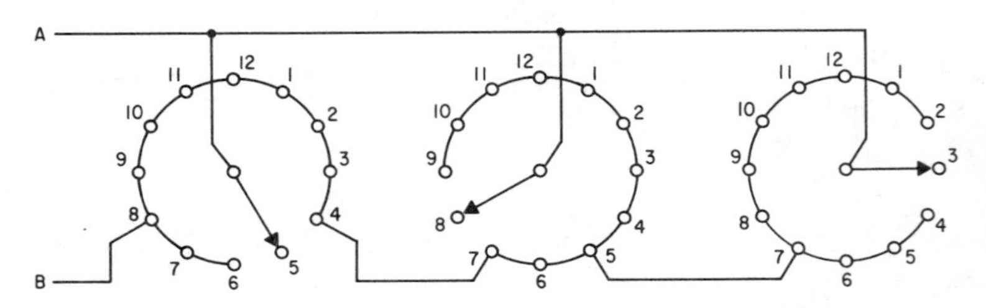

Figure 16-2 *Phone Lock Schematic*

Construction

Fabricating the phone lock is very simple. Purchase three (or as many as you wish) rotary switches. They should have about ten or twelve individual positions. Such a switch is designated as a single pole ten throw (SP10T) or single pole twelve throw (SP12T) switch. Mount the switches in a small bakelite case or any other convenient enclosure. Now, connect all of the centerposts of the three switches together. This will be wire A.

Determine what combination you want to use for the lock and short all the contacts on each switch together except the contact to the number you have chosen. That is, if you have decided that the combination you want is 583, on the first switch leave the terminal for position 5 empty and short all the remaining ones on that switch together. The same is done for terminal 8 on the second switch, and terminal 3 on the third switch.

Now connect all of the shorted terminals on the three switches together. This will be wire B. To check the circuit, place an ohmmeter or continuity tester across A and B. Dial in the combination. The ohmmeter or continuity tester should indicate an open circuit. If it doesn't you have connected the switches properly. If it does, turn one switch to another number. A short circuit should immediately be indicated.

Attach a piece of lamp cord, or just two individual wires, to A and B that are long enough to reach your phone. Fasten the case closed.

Installation and Operation

The phone lock can be attached directly to the telephone with a piece of double-sided tape, or can be installed at some remote point to prevent tampering.

To connect the device to the phone, remove the case by unscrewing the two screws located on the bottom of the telephone. Next locate the network block (see Fig. 1-1) which is the block inside the phone with all the screw terminals on top of it.

For telephones manufactured by Western Electric, ITT, or Bell Telephone, locate the green and blue wires coming from the dial to the terminal block. Connect lead A from the phone lock to the green terminal, which is usually designated RR on the terminal block. Wire B from the phone lock should be connected to the blue terminal, usually designated F.

For telephones manufactured by Automatic Electric or General Telephone, locate the blue and yellow wires coming from the dial to the terminal block. Connect lead A from the phone lock to the blue terminal, which is usually designated No. 1 on the terminal block. Wire B should be connected to the yellow terminal, which is usually designated as No. 11.

On some telephones the dial may have to be removed to get to the terminal block. This can be done by simply pressing down on the dial and pushing it towards the front of the phone.

Once the phone-lock leads have been connected, route the wires out of the phone and replace the cover. It is now ready to use. To place a call, set the combination on the phone lock and dial as you normally do.

To prevent unauthorized use of your telephone, simply turn each switch so that it is set on any number but the number in the combination. Now only incoming calls can be received. No outgoing calls can be made.

By the way, since the phone lock connects directly to the dial switch, it can be used to secure all the lines of a multiline phone, as well as single line ones.

One more important point, if you have one or more extension phones in the house, each one must have a phone lock on it to effectively secure your phone system.

17. Hold Button

The telephone company has come up with some very nice circuits that make using the telephone a little more convenient and helpful. Unfortunately, not all of these conveniences are always passed on to the average user. An example of this is the HOLD button found on multiline commercial phones. Wouldn't it be great if you could put the person you were talking to on HOLD for a minute while talking to someone else? You wouldn't have to worry about covering the mouthpiece on the phone with your hand.

Or maybe you picked up the telephone in the bedroom, but the conversation you're having requires that you move to another phone where it is more convenient to take notes. Instead of laying the phone down on the table moving to your other phone and continuing, and then worrying about remembering to put your other phone back on the hook, wouldn't it be wonderful to put the party on hold, hang up your telephone, move to another location, and pick up there?

But alas, HOLD buttons are not available on private single line phones. For a few dollars worth of components however, you can add your own HOLD button to your phone.

This project requires a direct connection to the telephone line. Since most telephone companies object to foreign attachments to their lines because of the fear that voltages might be introduced causing harm either to personnel or to the equipment, make sure you check with your local telephone company before making any direct line connections.

About the Circuit

The hold circuit is a relatively simple one and requires only six components. The circuit, which can be built in a little plastic box or even mounted right in the telephone, gets connected across the two telephone lines in parallel (Fig. 17-1).

To put a telephone on hold, it is necessary to raise the resistance across the line while it is in use. This prevents the line from releasing when you hang up the receiver. We could of course simply use a switch and a resistor to do this, but then you'd have to remember to turn the switch off, or you might forget that you had someone on HOLD.

To eliminate these problems we use a pushbutton, an SCR and a light-emitting diode (LED). The resistance R1 we use as the holding resistance is about 1200 ohms. When the telephone is switched from the operating mode to the hold mode, the resistance across the line changes from the 600 ohms of the off-hook telephone, to the 1200 ohms of the holding resistor.

In operation, when the telephone is off-the-hook, the voltage present on the line is about 5 volts. When the phone is not on HOLD, the LED is normally off. When the pushbutton switch is depressed and the receiver is returned to its cradle, R1 and R2 form a momentary voltage divider that applies a portion of the voltage across the telephone line to the gate of the SCR. This triggers the SCR on.

As soon as the SCR triggers, the 1200-ohm resistor is placed across the line, putting the party on the other end of the line on hold and preventing a disconnect. In series with the 1200-ohm holding resistor is an LED. The current flowing through the 1200 ohms is about 40 mA, which is sufficient to cause the LED to emit light. The LED will stay lit as long as the party is on hold and serves as a useful reminder that someone is waiting for you on the telephone.

When any phone on the line is picked up and someone starts to speak into it, the voltage drops across the SCR and the line is released.

Construction

The hold circuit can either be built directly into the telephone, or in a separate small bakelite box. If the phone you're using is not owned by you, build the circuit in a separate box.

Since there are so few parts involved, you may choose to build this circuit on a piece of perf board.

There are a few things to watch out for when building this circuit. The first is polarity of the LED. The LED must be connected so that the anode of the device goes to the positive (usually green) wire in your telephone. The next thing to look out for is the polarity of D2. Be sure that the cathode of D2 goes to the negative (usually red) wire of the telephone.

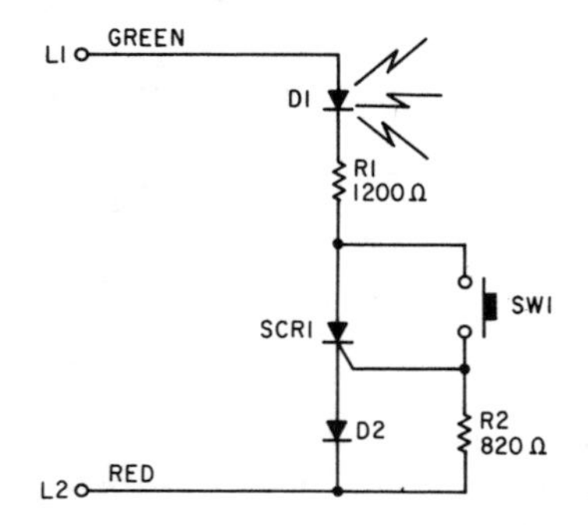

Figure 17-1 *Hold Circuit Schematic*

Installation and Operation

If the hold circuit is built into the telephone, all connections can be made to the terminal block inside the phone.

If it is being built as a separate unit, connections can be made either directly to the telephone junction box, or to a telephone plug or jack-in-a-plug.

To test the unit, call a friend. After he answers, tell him you are going to put him on HOLD for a minute and not to hang up. Then, while pressing the pushbutton, hang up the phone. As soon as the phone is on the hook, you may let go of the pushbutton. The LED should remain lit. If it does not, check to see if it is connected properly. Also see if the party holding is still on the line. If he is and you've checked the LED for correct polarity, it is possible that that LED is bad. If the LED is on then everything should be all right. To take the party off HOLD, it is simply necessary to pick up the receiver. The LED should go dim and then extinguish.

Parts List

D1 — light emitting diode
D2 — 1N914
R1 — 1200 ohms
R2 — 820 ohms
SCR1 — GE 106Y
SW1 — spst momentary contact pushbutton

18. Telecorder

At one time or another most of us have had the need to record a telephone conversation. To do so you've probably had to hunt around for a telephone pickup coil, attach it to the phone and your recorder, hope that the batteries in your tape recorder would last and then try not to get tangled up in the wires and accidentally pull the coil away from the telephone.

With a home built telecorder you can eliminate recording problems. For less than the cost of a commercial unit that does not contain its own power supply, you can build this device which will automatically record all incoming and outgoing calls from your phone. It works with your cassette recorder and any telephone or extension.

The telecorder contains a built-in regulated power supply which can be used to power the recorder and save batteries. This is particularly important in continuous monitoring applications.

About the Circuit

The heart of the telecorder is the TEL 100 telephone sensing module (Fig. 18-1). This module is the element that interfaces the phone line with the recorder. When connected to the red and green wires of the telephone, it senses the voltage across these two wires and produces a switching signal that activates a relay connected

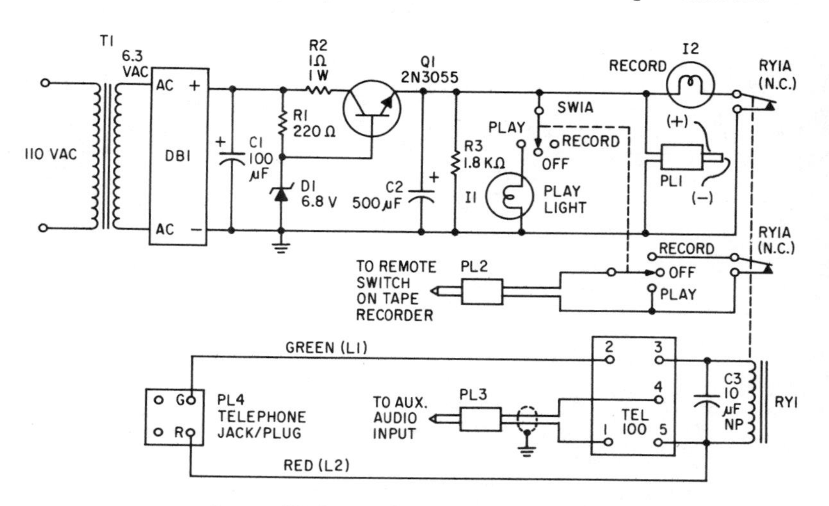

Figure 18-1 *Telecorder Schematic*

across terminals 3 and 5. A relay switching signal is produced every time the telephone receiver is lifted off the hook.

The TEL 100 module also isolates the recorder from the phone line and protects the input of the tape recorder from damage that might be caused by the 90V ringing signal. A 10 µF non-polarized capacitor is placed across the relay to keep the ring signal from affecting it.

The audio signal from the phone line is fed through the module into the auxilliary audio input jack of the cassette recorder. The operation of the tape recorder is controlled by the relay, whose normally-closed contacts are connected to the remote switch jack on the recorder.

The design of the power supply portion of the telecorder is relatively straightforward. The ac line voltage is stepped-down and rectified, and then applied to a regulating circuit. The output voltage of the supply is determined by the voltage across zener diode D1 minus the 0.7V drop across transistor Q1. If your recorder requires a 7.5V supply, substitute an 8.2V zener diode for the 6.8V one specified.

Parts List

C1 — 100 µF 16wVdc electrolytic	Q1 — 2N3055
C2 — 500 µF 16wVdc electrolytic	R1 — 220 ohms
C3 — 10 µF 50wVdc non-polarized	R2 — 1 ohm 1 watt
D1 — 6.8V zener diode	R3 — 1.8k ohms
DB1 — diode bridge, 50V piv	RY1 — 24V dpdt relay, coil resistance 2000 ohms
I1 — 6V lamp	SW1 — dpdt center off switch
I2 — 6V lamp	T1 — power transformer 115V primary, 6.3V secondary
PL1 — coaxial power plug	TEL 100 module (Fig. 18-2).
PL2 — subminiature phone plug	
PL3 — miniature phone plug	
PL4 — telephone jack-in-a-plug	

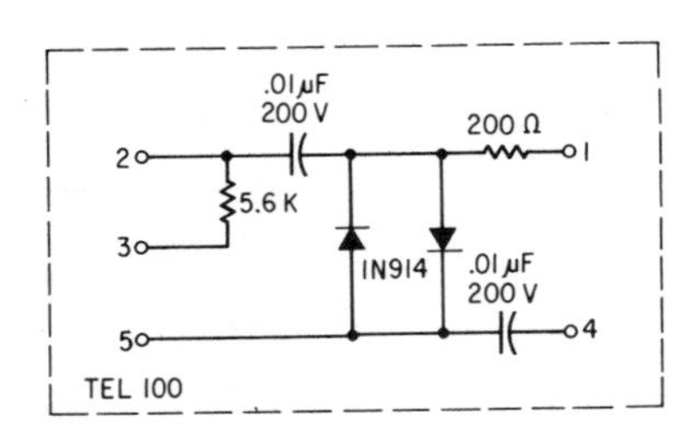

Figure 18-2 *Telecorder Module*

Construction

The circuit can be easily assembled on a perforated board. Except for the two panel lights and the control switch, all components are mounted directly on the circuit board. When installing the semiconductors be sure to observe the proper polarities and heat sink their leads while soldering.

In the prototype the circuit board was mounted in a 6¼" × 3⅝" × 2" plastic utility box. Three holes are drilled in the box to accommodate wires going to and from the telecorder. Be sure to line these holes with rubber grommets to prevent frayed wires.

While the layout of the circuit is not critical, it is important that the audio input lead to the tape recorder be shielded to prevent 60 Hz hum from being picked up. A two-conductor shielded cable is recommended with the shield connected to ground at the pc board. Connections to PL2 and PL3 are not critical in that any wire can be connected to the tip or the body of the plug.

Installation and Operation

The telecorder (see Fig. 18-2) can be connected to any telephone or directly to the telephone junction box. But the easiest way to hook it up is to use a telephone jack/plug which is available from most electronic parts suppliers. This device fits in between a standard telephone plug and a standard jack. Two of the four terminals on the jack/plug are marked R and G (for red and green). The L2 wire from the telecorder is connected to the R terminal and the L1 wire is connected to the G terminal. The jack/plug is then inserted into the telephone jack and the plug on the telephone is inserted into the telecorder's jack/plug.

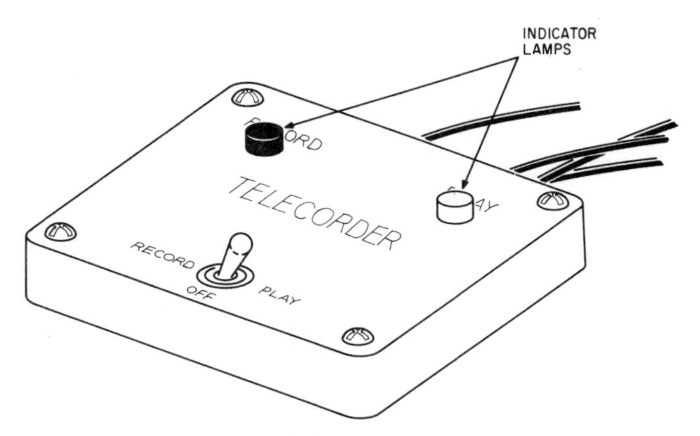

Figure 18-3 *Telecorder*

When the jack/plug is inserted into the telephone jack, you should hear the relay click. If you do not, check to make sure the wires are properly connected to the jack/plug. Now with the switch in the center (off) position, plug in the ac line cord. None of the panel lights should go on. If the red light is on, the relay isn't working and the unit is probably incorrectly connected to the phone line. If the play light is on, the wiring to the switch is at fault.

Now put the switch in the PLAY position. The PLAY light should light up and the recorder should be operational. If it is, put the switch in the RECORD position. The PLAY light should go out and the recorder should stop. At this point the RECORD light should not be on. Lift the telephone receiver off the hook. The RECORD light should now go on and the recorder should now be taping anything that is heard in the telephone receiver. When you replace the receiver in its cradle, the light should extinguish and the recorder should stop.

For a continuous monitoring operation, the telecorder is connected to both the recorder and phone line, and both units are placed in the RECORD mode.

It should be noted that although it is not illegal to connect privately-owned equipment to the telephone line (due to the 1968 Carterfone Decision), in some areas of the country it is against internal phone company regulations. In those areas, it is necessary to place a recorder coupler between the phone line and the equipment to be connected to it, for the device to be strictly legal. If you want to be sure if a recorder coupler is required in your area it is best to check with your local phone company.